BIRDS of THAILAND

Michael Webster and Chew Yen Fook

POCKET PHOTO GUIDE

BLOOMSBURY
LONDON • OXFORD • NEW YORK • NEW DELHI • SYDNEY

Bloomsbury Natural History
An imprint of Bloomsbury Publishing Plc

50 Bedford Square
London
WC1B 3DP
UK

1385 Broadway
New York
NY 10018
USA

www.bloomsbury.com

First published by New Holland UK Ltd, 1998 as *A Photographic Guide to Birds of Thailand*
This edition first published by Bloomsbury, 2016

British Library Cataloguing-in-Publication Data
A catalogue record for this book is available from the British Library.

Library of Congress Cataloguing-in-Publication data has been applied for.

ISBN: PB: 978-1-4729-3792-6
ePDF: 978-1-4729-3795-7
ePub: 978-1-4729-3793-3

2 4 6 8 10 9 7 5 3 1

Designed and typeset in UK by Susan McIntyre
Printed in China

CONTENTS

INTRODUCTION

Nobody can write about the birds of Thailand without recognizing a tremendous debt to two people in particular: the late Dr Boonsong Lekagul, the force behind Thailand's conservation movement, and the author of books on birds, mammals and butterflies, and Philip D. Round, who, without question, knows more about birds in Thailand than anybody else. The standard field guide, *A Guide to the Birds of Thailand*, was written by Philip Round on the foundations laid by Dr Boonsong in two earlier versions.

In the present book, that guide has been used extensively as a basis for distribution details, and for field information about birds which the author himself has not seen. It is in addition the source of all information on calls (with Philip Round's permission), as it was thought sensible to standardize descriptions where possible. English and scientific names also follow Lekagul and Round in principle, though some have been updated in accordance with the most recent (2008) *A Field Guide to the Birds of South-East Asia* by Craig Robson.

Lekagul and Round list 915 species which have been recorded in Thailand, and a few more have been added to the list since their guide was published. This book illustrates 252 species, slightly more than a quarter of the total, and many others are referred to in the text. All the illustrations are real-life photographs showing the 'jizz' of the birds as they are likely to appear in the field. While this range of species should cover the vast majority of species which the average visitor can expect to see, there will no doubt be some for which the dedicated birdwatcher may need the more comprehensive book.

Many species have undoubtedly been lost through the wholesale destruction of habitat, especially the forests, during the past fifty years, and many are endangered as suitable habitats continue to shrink.

HOW TO USE THIS BOOK

This guide has been designed for ease of use and quick reference when birdwatching in Thailand. The introductory section contains basic necessary information, including a map and description of the different regions of the country, and details of some important birdwatching areas. This is followed by the 252 species descriptions and photographs.

To identify a bird, using the book, first check the key to the corner tabs on page 7 to see which of the silhouettes most closely matches what you have seen. This will give you the family or group of families to which the bird belongs, which you can then look up in the index. Most of the photographs are of the male of the species; major differences of female plumage are mentioned in the text. Waders have generally been photographed in their winter plumage, as these birds are unlikely to be seen in Thailand in their summer plumage except for a short period towards the end of the spring migration.

Where possible, the photographs have been chosen so that they show the distinctive characteristics of the species. These may include an indication of habitat and of 'jizz' (the bird's general shape, the way it stands/perches), and of course its colours. This will not always, however, enable you to match the bird you see with the photograph.

The accompanying text summarizes the additional information which you will need for identification. The description for each species begins with the common name, the scientific name, and the length of the bird in centimetres. The first sentence gives a quick introduction to the species, generally referring to the most striking plumage feature or most distinctive habit. The next few sentences give the general identification features, including call and behaviour where appropriate, and information on how to distinguish the bird from similar species. The description concludes with a note on the bird's status, whether resident or migrant, its preferred habitat, and the region or regions in which it occurs.

The text uses a certain amount of technical language some of which is illustrated in the following diagram.

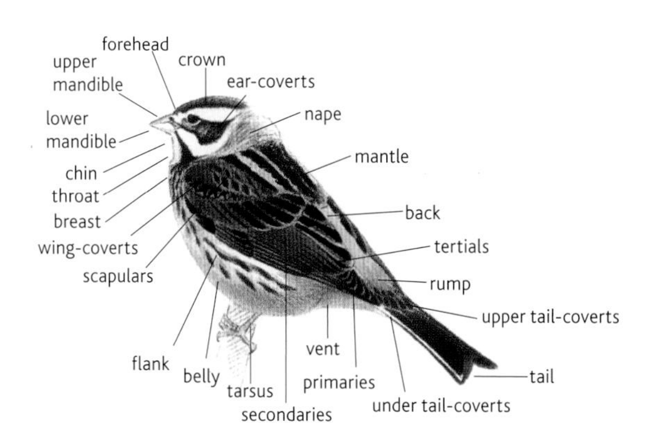

Specialized vocabulary used in the text but not illustrated in the diagram is explained in the following glossary.

GLOSSARY

bib Patch covering the throat and the upper half of the breast, generally sharply differentiated from the rest of the underparts
bird wave In montane jungle, a number of species travelling together in a feeding flock
canopy The largely unbroken layer of branches forming the tops of the trees in a forest
carpal joint The bend of the wing
casque A horn-like protuberance on the top of the bill
crest A tuft of feathers on top of the head
dimorphic Occurring in two different forms (e.g. sexually dimorphic, when the sexes differ in plumage)
drum Tap rapidly on a tree trunk
ear-tufts Tufts of feathers resembling ears (as on some owls)
eye-ring A ring of feathers encircling the eye
facial disc A clearly marked circular 'face' (owls)

facial skin Bare skin around the base of the bill
feral Having become wild, or descended from birds which have become wild
forewing The front part of the inner half of the wing
frontal shield A hard (bony) unfeathered patch on the forehead
genus A group of closely related species
gorget A half-collar roughly between the throat and the breast
graduated Of decreasing lengths (on some birds the central tail feathers are longest, and each successive pair, working outwards, is shorter)
hackles Feathers on the neck which can be raised
hover Remain motionless in the air, with wings beating, but without forward movement
lore Small area of feathers between the bill and the eye
mandible The upper or lower half of the bill
mask A (usually black) patch on the face, resembling a mask
moustache/moustachial stripe A line running downwards at approximately a 45° angle from the base of the bill
mudflats The wet area of mud left at low tide
orbital ring An unfeathered bare ring encircling the eye
plume A tuft of feathers on head, breast or lower back
post-ocular Behind the eye
roost A regular sleeping-place
subspecies A geographically separate form of a bird which is still part of the same species (i.e. can interbreed with it and produce fertile offspring)
subterminal Just before the tip or end
superciliary/supercilium A stripe over the eye
terminal At the tip or end
trailing edge The rear edge of the wing
underwing-coverts Small feathers covering the base of the primaries and secondaries on the underside of the wing
wattle A bare fleshy protuberance on the head
wedge-shaped Of tail feathers, coming to a point in the middle when spread, instead of forming a curve or straight edge
wing-coverts Small feathers covering the base of the primaries and secondaries on the upperside of the wing
wing-linings Same as underwing-coverts
wingbar A line of colour (or white) along the wing

KEY TO COLOURED TABS

Grebes & cormorants,

Herons, egrets & storks

Ducks

Raptors

Gamebirds

Rails & allies

Waders

Terns

Pigeons, doves & parrots

Cuckoos & relatives

Owls

Trogons

Kingfishers

Bee-eaters, rollers & hoopoes

Hornbills

Barbets & woodpeckers

Broadbills & pittas

Swifts & swallows

Pipits & wagtails

Cuckoo-shrikes

Minivets & leafbirds

Bulbuls

Drongos & orioles

Crows

Tits

Babblers & allies

Laughing-thrushes

Flycatchers & relatives

Warblers & tailorbirds

Thrushes & relatives

Shrikes

Starlings & mynas

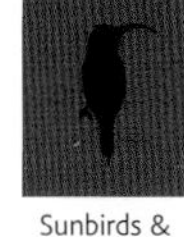
Sunbirds & spiderhunters

Flowerpeckers

White-eyes, sparrows & relatives

HABITATS

The large number of species recorded reflects Thailand's wide variety of habitats, ranging from coastal mudflats to mountains above 2000m. These have been separated into regions as follows (localities likely to be visited by tourists are listed in brackets).

1. North Largely montane forests of the northern part of the country, which have a long dry season from November through to May. These have been extensively logged, but patches of primary forest remain in some areas. This area extends south and east to Nam Nao Forest Reserve. (Doi Suthep, Doi Inthanon, Nam Nao.)

2. West Montane and submontane forests down the western border, from Mae Hong Son south to Kaeng Krachan. Similar to the north, but often less disturbed, and containing the biggest reserve areas. Except in the northern parts, most of this area is not so mountainous. (Huai Kha Khaeng.)

3. Central plains Flat, mostly rice-growing country north of the capital. The country's biggest marsh, a major wintering area for waterbirds, is in this area. (Bung Boraphet.)

4. North-east Largely dry, arid inland provinces, with little remaining forest; the well-known forest reserve Khao Yai is on the southern border of this area. The eastern boundary of this region is formed by the River Mekong, which still supports a few riverine species. (Khao Yai.)

5. South-east A relatively small area south of Khao Yai and east of Bangkok. Rainforest, with a higher rainfall than most other parts of Thailand.

6. Peninsula The long narrow isthmus south of Bangkok extending to the Malaysian border. Many species occur only in this region. A variety of habitats may be found here, but most of the lowland forest has been destroyed. (Khao Sam Roi Yot, Krabi.)

Most visitors to Thailand are likely to stay in the capital, or in a limited number of provincial centres. From Bangkok, Khao Yai is three hours' drive away (traffic permitting); Bung Boraphet is a similar distance. Chiangmai and Chiangrai are the centres of tourism for the north; the former is within easy reach of Chiang Dao (Doi Luang) and Doi Inthanon, while Doi Suthep is just outside the city. Hua Hin is an hour's drive away from Khao Sam Roi Yot.

BIRDWATCHING IN THAILAND

Most of the recommended localities are for forest birds, exceptions being Bung Boraphet and Khao Sam Roi Yot. The northern forests are best visited during the long dry season, when the leeches and mosquitoes are not in evidence. You can expect to walk quite long distances, but the climate is fairly equable. Start early in the day, at dawn if practicable. Carry plenty of water, as you may not be able to replenish supplies en route; do not drink from streams, since they may have been contaminated by villagers further up the mountain. Similarly, you should carry your food for the day unless you are sure to be able to reach a restaurant or food stall. Wear dull-coloured clothing, and carry a lightweight plastic rain-cape if rain seems likely (this is unnecessary in the north in the winter months).

During the rainy season (May to October) in the north, and throughout the year in all other parts of the country, take precautions against mosquitoes and leeches. Some areas, especially in the west, are still malarial; preventive drugs do not work very well, and the best precaution you can take is to avoid being bitten. This means long sleeves, trousers (not shorts), and plenty of mosquito repellent. Such precautions will also help you against leeches; keep in the centre of the track, and avoid wet grassy areas in the forest.

As with forest birdwatching everywhere, you have to rely on your ears to locate most of the birds. Do not expect to identify everything; do follow up calls whenever possible. Many forest species, especially in the hills, travel in mixed flocks known as 'bird waves'. This means that several species (of insect-eaters) associate in a feeding flock. When you come across such a flock, try to identify as many birds as you can; do not assume that the ones you have not yet identified are the same as those you have already spotted.

In Bung Boraphet and Khao Sam Roi Yot (except in the little wood near the Visitors Centre) you will not find much shade, so it is necessary to wear some kind of headgear; sunburn is also a risk if you are susceptible to it.

You will need a good pair of binoculars in all areas, and a telescope if you are watching waders, ducks or seabirds. A telescope is also useful in the forests, but is not essential if you feel that it is too heavy to carry. Take a notebook and pencil or ballpoint pen to note down what you see, together with descriptions of unusual or unfamiliar species. Get used to writing descriptions of birds. Sound-recorders and cameras are optional; you may prefer to travel light. Birdwatchers are advised not to use a recorder to 'call birds in'; the birds may respond to a recording of their territorial calls, but this use of recordings may also cause major disturbance to their breeding behaviour.

LOCALITIES

The localities marked on the map and listed below are among the best birdwatching sites in Thailand; there are many others! Sites which are difficult to access are not included.

Chiang Saen Paddy, marshland, and a lake along the Mekong in the extreme north. Good variety of species in winter. About an hour by car from Chiang Rai.

Doi Angkhang A largely deforested mountain in the north; some good montane species. About two and a half hours by car from Chiang Mai.

Doi Chiang Dao One of Thailand's highest mountains; good variety of forest species. About an hour's drive from Chiang Mai.

Doi Suthep/Pui Good range of northern montane species; on the outskirts of Chiang Mai.

Doi Inthanon Thailand's highest mountain, and the best locality for forest birds in the north. About two hours' drive from Chiang Mai; road goes to the top.

Nam Nao On the main road from Pitsanulok to Khon Kaen. Good variety of forest species; low hills. Overnight stay is usually possible; bungalows/tents may be hired.

Phu Luang A montane site in the north-east (Loei Province). Access by car from Loei.

Bung Boraphet A large wetland in the central plains, important for breeding and wintering waterfowl. Three hours by car from Bangkok; just outside the town of Nakhon Sawan.

Khao Yai Lowland and submontane forest; one of the best and most accessible localities. About three hours' drive from Bangkok; accommodation available outside the park.

Khao Ang Ru Nai and **Khao Sai Dao** Lowland and montane sites respectively. The best localities in the south-east; access from Chanthaburi.

Huai Kha Khaeng This, with the adjacent forests of Umphang and Thung Yai Naresuan, forms the largest remaining block of forest in Thailand; excellent for lowland and submontane species. Access from Nakhon Sawan.

Gulf of Thailand Mudflats for migrant and wintering shorebirds; best access from Bangkok at Bang Poo, and between the Tachin and Mae Klong Rivers.

Kaeng Krachan The biggest single forest reserve, still not well known. About three hours' drive from Bangkok.

Map of Thailand showing birdwatching localities

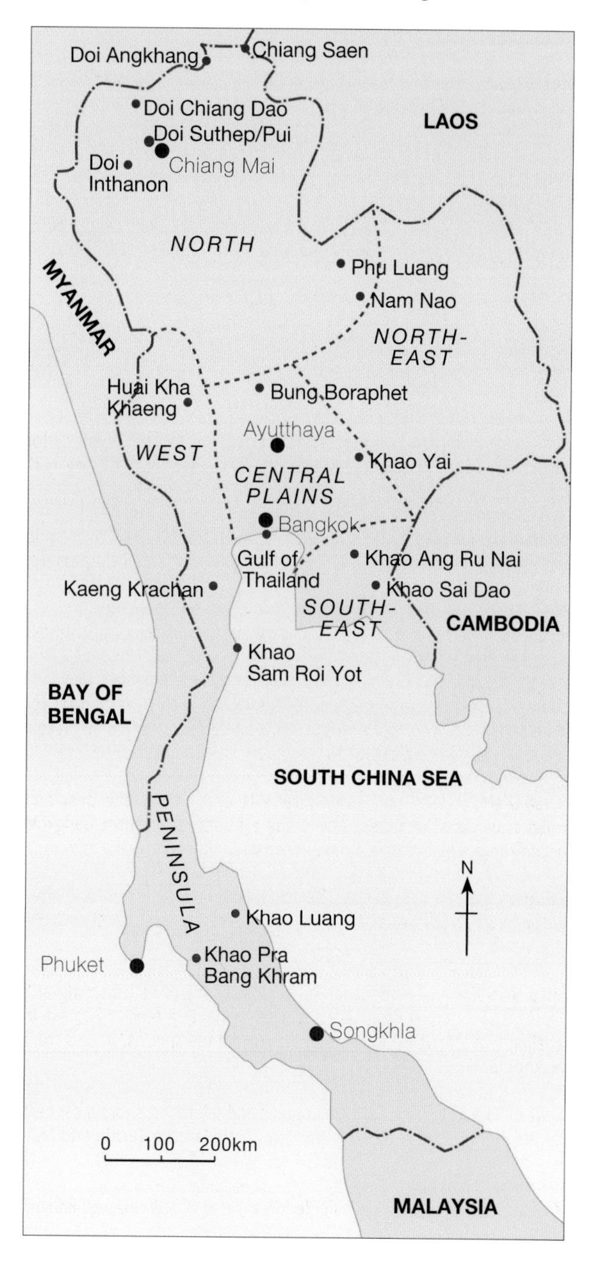

Khao Sam Roi Yot Freshwater marsh and mudflats. An hour's drive from Hua Hin; accommodation in the reserve.

Khao Luang The best montane forest left in the Peninsula. Access from Nakhon Si Thammarat.

Khao Pra Bang Khram Lowland forest, the best remaining area in Peninsular Thailand. Near Krabi, which is a good centre for birding in the south.

BIRDWATCHING SEASONS

In this book, birds are classified under the following headings:

1. *Resident* Birds so described remain in the country throughout the year, and may be seen at any season.

2. *Summer visitor* These birds arrive in Thailand in April/May, and stay until August/September; they breed during this period. A few come here to breed during the dry season (e.g. Asian Openbill).

3. *Winter visitor* These are Palearctic (northern) species which come to Thailand for the northern winter, roughly from late September until the end of March, before returning north again.

4. *Passage migrant* Birds which pass through Thailand on their migration. In September and October, many species pass through on their way south (notably the waders, en route to Indonesia and Australia), and the same species return from late March until May. The timing of this migration is influenced by various factors, some of them regular, such as length of daylight and seasonal weather patterns, and some of them irregular, such as temporary weather conditions.

5. *Vagrant* This category covers species which have been recorded only a few times in Thailand. Genuine vagrants may be brought here by extreme weather conditions. Thailand is, however, a large country, and many places are rarely visited by birdwatchers; new discoveries are therefore still possible, and new species are being added to the list almost every year.

FURTHER READING

The following books will also help you find and identify Thai birds.

King, B., Woodcock, M. and Dickinson, E.C. 1975 (and subsequent editions). *A Field Guide to the Birds of South-East Asia*. London: Collins.
Covers the birds of the whole of South-east Asia, from Burma eastwards to Vietnam and south to Singapore. Excellent for field identification, if a little confusing because of the large number of species covered. Well illustrated.

Robson, C. 2008. *A Field Guide to the Birds of South-East Asia*. London: New Holland Publishers.
Fully updated and illustrated edition of the definitive modern guide to the region.

Robson, C. 2002. *A Field Guide to the Birds of Thailand*. London: New Holland Publishers.
The most up-to-date and comprehensive book on the birds of Thailand, including full-colour plates and distribution maps for individual species.

Round, P.D. 1988. Resident *Forest Birds in Thailand: their status and distribution*. Cambridge: International Council for Bird Preservation.
Now somewhat out of date, but still the best handy guide as to what to look for in specific forested areas.

Photographic Acknowledgements

All the photographs in this book were taken by Chew Yen Fook with the exception of the following: David Bakewell (78 bottom); Lauri Maenpaa (98 bottom); Ken Scriven (68 bottom); Morten Strange (34 top, 55 bottom, 75 top); Ray Tipper (14 top, 21 bottom, 37 bottom, 38 top, 41 top, 43 top, 62 bottom, 77 top & bottom, 78 top, 86 top, 108 top, 109 top & bottom, 111 top, 119 top, 125 top); Bob Watts (63 bottom, 90 top); Windrush Photos: Goran Ekström (66 top) and David Tipling (36 top, 75 bottom).

Author's Acknowledgements

Michael Webster would like to thank Phil Round and Kamol Kamolphalin for their help and support throughout the project.

LITTLE GREBE *Tachybaptus ruficollis* 25cm

A small, dumpy waterbird which characteristically dives when feeding and when alarmed, surfacing again a few metres away. Dark brownish-grey above, pale brown below. The sides of the head and neck are dark chestnut in the breeding season. At close range, there is a bright yellow spot at the base of the bill. Much more secretive in the breeding season. Seldom flies, though it may patter across the surface of the water. It breeds at the edges of lakes or marshes, making its nest on a floating platform. Common resident.

INDIAN CORMORANT *Phalacrocorax fuscicollis* 64cm

The longer-billed of the two cormorants regularly found in Thailand. Length of bill equals approximately length of head. The adult in summer has a white ear patch and yellow throat pouch, but is otherwise all glossy black. Wintering birds have a whitish throat, and immatures are a dirty white below. Habitat similar to that of Little Cormorant, but has a particular preference for the larger rivers. Uncommon resident in and near the lower reaches of the Chao Phraya River.

LITTLE CORMORANT *Phalacrocorax niger* 52cm

Large and black, with a white patch on the throat in winter. This is the common cormorant of the central plains. It has a much shorter bill than the Indian cormorant; the length of the bill equals approximately half the length of the head. It lacks the white ear-patch. When perched, cormorants may be confirmed by their diagnostic stance, with the body often held at 45 degrees, or the wings 'hanging out to dry'. Resident on inland waters, including flooded paddyfields, mangroves and coastal areas.

ORIENTAL DARTER *Anhinga melanogaster* 91cm

A peculiar-looking bird with distinctive long, snaky neck; when the bird swims, the body is submerged. The plumage is generally black, though the head and neck are brown, and there are streaks of white on the upperparts and the sides of the neck. In soaring flight, the neck is outstretched but kinked, and the tail is long. This spectacular resident of lakes and marshes is now likely to be seen only in remote areas, and may no longer breed. Probably most frequently seen along the western border.

GREY HERON *Ardea cinerea* 102cm

A large and fairly distinctive pale grey heron, with whitish head and neck and a black drooping crest. In flight appears grey and black, the black flight feathers contrasting with the grey wing-coverts and body feathers. On the ground, the long neck may be extended or drawn back in hunched posture; in flight, the neck is folded back to the body and the legs trail. A winter visitor to lowland areas throughout the country, both at inland wetlands and at coastal marshes and mudflats.

PURPLE HERON *Ardea purpurea* 97cm

This rather attractive bird is distinguished from the Grey Heron by the comparative lack of contrast in the flight pattern. At rest, the body is purplish-grey, the neck a bright rufous colour. It prefers marshy reedbeds, from which the long snake-like neck may be seen sticking up; it rarely hunches up like the Grey Heron. In flight, the neck is folded back as in the Grey Heron. It appears all dark, with little contrast. A winter visitor to lowland marshy areas throughout the country; a few birds breed (e.g. in Khao Sam Roi Yot).

CHINESE POND-HERON *Ardeola bacchus* 46cm

Winter (above); breeding plumage (right)

The common small heron of coastal and inland waters. In flight it appears white with a dark body, but most of the white disappears when it lands (as the wings are folded). In winter it is light brown and streaked, with white wings. In breeding plumage, the head and breast become deep chestnut, and the back blackish. Prefers freshwater marshes and paddyfields, as well as coastal mangroves and mudflats. This species is a common winter visitor. The Javan Pond-Heron *Ardeola speciosa*, which breeds in the central plains, is distinguished in breeding plumage by the pale buff or whitish head and throat.

CATTLE EGRET *Bubulcus ibis* 51cm

This white heron-like bird is commonly seen associating with cattle or buffaloes, preferring grassland to the marshes or mudflats. Distinguished from other egrets by the heavy yellow bill and shorter neck. In breeding plumage the head and neck are light orange, bill and legs pinkish. Common resident in most lowland areas. Other yellow-billed egrets include the Great Egret (much larger: 90cm) and the Intermediate Egret *Egretta intermedia* (71cm; taller, more slender, with usually a black tip to the bill).

GREAT EGRET *Ardea alba* 90cm

The largest of all the egrets, with a very long and slender neck, this handsome species has a completely white plumage at all times of year. The bill is yellow in non-breeding plumage and black in breeding plumage. The legs are blackish (usually with reddish upper half early in the breeding season). It does not normally flock like the Little Egret, which also occurs in the same habitats, although it is not unusual for there to be many in a small area. A common resident of marshland, but also frequents the coastal mudflats.

LITTLE EGRET *Egretta garzetta* 61cm

The only small white heron with a slender black bill (the much larger Great Egret has a black bill only in breeding plumage). Delicate head, breast and back plumes in breeding plumage. The black legs and contrasting bright yellow feet are also distinctive. Frequently seen in large flocks, often with other herons and other large wading birds. Found on most types of wetland, including lakes, marshes and fish ponds, but less commonly on the coast. Common resident and winter visitor in most of the country.

LITTLE HERON *Butorides striata* 46cm

A small, often inconspicuous, greenish-grey heron found among mangroves; in winter, it may be flushed along forest streams. The sides of the neck are greyish; the legs and feet are yellow. Common resident in the south, and a winter visitor throughout the country, except the north-east. A somewhat similar species, the Black Bittern *Dupetor flavicollis*, which looks longer and less compact in flight, has a buff patch on the neck, and dark legs, and breeds commonly in central Thailand.

BLACK-CROWNED NIGHT-HERON *Nycticorax nycticorax* 61cm

An attractive and boldly marked heron. Adults show a distinctive pattern of black back and crown, and pale grey wings; underparts are greyish-white. The legs are yellow. Immature birds are brown, speckled white. This species is gregarious and largely nocturnal, except during the breeding season. Daytime roosts can often be located by the chorus of hoarse farmyard noises from within a mangrove swamp or similar site. Common resident in the central plains, and a winter visitor in some areas.

YELLOW BITTERN *Ixobrychus sinensis* 38cm

The pale brown wings contrasting with black primaries are diagnostic when this bird is in flight; rarely seen at rest. This tiny heron breeds in marshland, preferring the taller reedbeds. A common resident and winter visitor in much of the country (except the north-east and east). The Cinnamon Bittern *Ixobrychus cinnamomeus* is bright chestnut (adult male), darker than Yellow Bittern (other plumages), and occurs more frequently in grassy areas than the Yellow Bittern.

ASIAN OPENBILL *Anastomus oscitans* 81cm

A large black and white bird which flies with its neck outstretched. This is the only remaining stork which can be seen with any regularity in Thailand. The primaries and rear wing are black, as are the tail and the scapulars; the rest of the plumage is dull white. When the bill is closed there is a gap between the upper and lower mandibles, from which the bird gets its name. Occurs in marshes and paddyfields. Breeds in a few colonies in the central plains (November to April), most migrating to Indian subcontinent for remainder of year.

WOOLLY-NECKED STORK *Ciconia episcopus* 91cm

This predominantly glossy black bird has a white neck, lower belly and undertail-coverts. The bill is black and the legs red. Immatures show a similar pattern, but are browner. It is the smallest of the storks in Thailand, except for the commoner and easily recognized stork the Asian Openbill. Formerly a not uncommon resident in the marshlands and around pools in more open forests of the lowlands, it is now very rare, but most likely to be seen in the peninsula or the south-east.

NORTHERN PINTAIL *Anas acuta* 56cm

The long, slender neck and pointed tail make this one of the easiest ducks to recognize, both in flight and at rest. In the summer-plumaged male, the neck is chocolate-brown with a white line up the side; the body is mainly grey. The female is mottled brown, and is best told by shape (relatively long neck and short but pointed tail). In flight, it shows a narrow white trailing edge to the secondaries. Common winter visitor to lowland lakes and marshes.

GARGANEY *Anas querquedula* 41cm

The commonest wintering duck. The male is distinguished at close range by the broad white band over the eye; in flight, it shows a pale blue-grey forewing which contrasts with the dark remainder of the plumage. The female is separated with difficulty from the female Common Teal *Anas crecca* by the more striped head pattern, including a dark bar across the cheek. Common winter visitor, found on marshes, lakes, and near the sea coast, often in large flocks.

COMB DUCK *Sarkidiornis melanotos* 76cm (male)

The large size and dark back, with whitish neck and underparts, distinguish this species from the only other duck of similar size, the White-winged Duck. The male has a characteristic 'comb' on the bill. The female, which is significantly smaller than the male and lacks the comb, is similarly patterned, but somewhat duller. In flight, the wings are blackish above and below; the head appears small. A rare resident of marshy areas, mainly in the central plains area.

WHITE-WINGED DUCK *Asarcornis scutulata* 76cm

A large, goose-like forest duck, identified in flight by the pure white forewing. A dark grey-brown bird with whitish head and neck. It frequents streams, lakes and pools in thick forest, characteristically flighting to roost (honking as it goes) in the branches of a favourite tree. Its regular habits have made it an easy target for hunters, with the result that it is critically endangered. It now remains in only a few localities in Thailand. Resident.

LESSER WHISTLING-DUCK *Dendrocygna javanica* 41cm

A medium-sized, brown duck with broad, rather rounded wings. At close range, the plumage can be seen to be reddish-chestnut, with the head and neck pale buff. Blackish primaries and secondaries. The commonest breeding duck throughout Thailand wherever there are ponds and marshes. In the breeding season, it may be seen flighting to roost at dawn, generally in pairs; in winter, large flocks form, often in the same localities year after year. Common resident and winter visitor.

OSPREY *Pandion haliaetus* 55–61cm

An Osprey plunging into the sea for fish is one of the most dramatic sights in the bird world. Feeding exclusively on fish, it hovers over the water before diving for its prey. As it rises into the air, there is a characteristic stagger in its flight as it shakes off the water. No other large raptor regularly hovers over water. Dark brown above, white below, with a white head and broad black eye-stripe; some birds have a darker breast band. Uncommon winter visitor to coasts and lakes; a few remain in the summer but do not breed.

BLACK-SHOULDERED KITE *Elanus caeruleus* 28–35cm

This graceful hawk is frequently seen hovering over open country and rice paddies. Pale grey above, with a black shoulder patch; white below, with black underside to the primaries (and sometimes a dusky band along the secondaries). Glides with its wings held at a sharp upward angle ('a steep V'). This common resident is found throughout Thailand in open country, including cultivated areas, and up to 1500m in the mountains. In the majority of areas it is probably the commonest hawk over open country.

BLACK KITE *Milvus migrans* 61–66cm

A large, dark brown hawk characterized in flight by its shallowly forked tail (more straight-ended when fanned); and also by the paler patch at base of primaries on the underwing. When flying, it constantly angles its tail from side to side. Wintering birds form large roosts, above which tens or hundreds of individuals may be seen soaring. The wings are arched slightly when soaring. Winter visitor and passage migrant, mostly in open country and lowland areas; a few may be found breeding in the central plains.

BRAHMINY KITE *Haliastur indus* 43–51cm

A large hawk with very distinctive adult plumage: it is chestnut, with a white head and breast and with dark wingtips. The immature is easily confused with a Black Kite, but is somewhat smaller, has shorter wings, and the tail is rounded (not forked or straight-ended) when fanned. The wings are held out almost horizontally when it is soaring or gliding. Common resident in coastal areas (also found at a few localities inland), frequently scavenging for scraps floating on water's surface.

BLACK BAZA *Aviceda leuphotes* 33cm

The only medium-sized black and white hawk. Long, erect crest, visible only when the bird is at rest. In flight, the underside of the wings is mainly pale grey, with dark underwing-coverts and inner secondaries and black tips to primaries; tail pale grey. Flight flapping, like that of a crow. Often seen in flocks. Found in wooded areas throughout the country, except in the north-east, most commonly on passage and in winter; a few are resident. It sometimes enters orchards and larger gardens.

WHITE-BELLIED SEA-EAGLE *Haliaeetus leucogaster* 60–69cm

This magnificent bird may be seen soaring, with wings held in a marked V-shape, over coastal waters. The head, underbody and underwing-coverts, and outer half of the tail, are pure white; the rest of the plumage is dark grey. The tail is wedge-shaped. The wings narrow towards the tip. Immatures are mostly brown, variously marked, but show the same distinctive V-shaped wing posture in flight. Uncommon resident along the coasts, often nesting on small rocky islets, where its builds a huge nest of sticks.

CRESTED SERPENT-EAGLE *Spilornis cheela* 51–71cm

A large forest hawk, with broad, long, rounded wings, often seen soaring singly or in pairs. The loud ringing call can be heard for a considerable distance. When perched, it appears a deep maroon, flecked with white below, and the head appears wedge-shaped as a result of the crest. In flight, the adult (seen from below) shows a conspicuous black-bordered white bar along the full length of the wing, and a white central band on the blackish tail. Resident in all forested areas.

BLYTH'S HAWK-EAGLE *Nisaetus alboniger* 51–58cm

Hawk-eagles are large, broad-winged and relatively long-tailed hawks with most species also having a conspicuous crest. The adult of Blyth's Hawk-eagle can easily be identified in flight by the broad white band across a blackish tail. At rest, the face-mask and upperparts are blackish, with brown scaling on the wings, and the underparts white, with black-barred belly and black-streaked breast. Young birds are difficult to separate from other hawk-eagles in flight. Resident in the mountains of the peninsula.

WESTERN MARSH HARRIER *Circus aeruginosus* 48–56cm

The Western Marsh Harrier hunts by quartering low across marshlands, its wings held in a shallow V. The wings are broader and more buzzard-like than those of other harriers. Plumage varies with age and sex; most birds are dark brown, with some white on uppertail-coverts, and a barred tail. Older males become grey and blackish (never solid black as on Pied Harrier *C. melanoleucos*), and the head may be whitish. Similar to Eastern Marsh Harrier (*C. spilonotus*) but markings on head, neck and breast are browner and ear coverts are paler.

SILVER PHEASANT *Lophura nycthemera* 51–120cm

Male (left); female (above)

This splendid pheasant, commonly kept in captivity, is still found in the wild in Thailand. The male is whitish above, with delicate black V-shaped markings, and has a long, largely white tail; the underparts are black. It has bright red wattles and red legs. The female is brown, with blackish underparts scaled whitish. Feeding parties utter continuous soft mewing sounds. It often associates with the Red Junglefowl. Resident in the north of the country, where it frequents the evergreen forests of the hills.

CRESTED FIREBACK *Lophura ignita* 69cm

A handsome bird with a fairly prominent crest. The male is the darkest of all Thai pheasants, with its plumage being almost entirely blackish-blue; the rump is orange-red, and the central tail feathers and streaks on the flanks are white. Facial skin blue, legs and feet reddish. The female is dull brown, with white scale-like markings on the underparts. This spectacular species of pheasant survives only in parts of the peninsula, where it is a rare resident in evergreen forests, usually near streams.

Male (above); female (below)

SIAMESE FIREBACK *Lophura diardi* 60–82cm

The golden-yellow patch on the rump is most conspicuous during display. The remaining plumage of the male is dark grey, with white markings on the wings. Wattles and legs are red. The female is much more heavily marked than other female pheasants, with broad black and whitish-buff bars on the wings and tail. A few are resident in the forests of the north-east, central plains and the south-east, but the species has been much reduced by deforestation.

Male (above); female (below)

RED JUNGLEFOWL *Gallus gallus* 43–76cm

Male (above); female (below)

This ancestor of the domestic chicken is very similar to the domestic bantam, but can be distinguished by the largish white rump patch of the male, and by the slate-grey legs of both sexes. The characteristic chicken call, 'cock-a-doodle-doo', is slightly higher-pitched than that of the domestic chicken and stops abruptly, with the last syllable almost or completely omitted. Frequently perches in trees. Common resident, the commonest and most easily seen of all pheasants and partridges in forest areas.

Grey Peacock-pheasant Polyplectron bicalcaratum 56–76cm

Peacock-pheasants, argus pheasants and peafowl all have ocelli ('eyes') on the wing feathers and/or the tail feathers. This species is grey-brown all over, with conspicuous white-edged violet to green 'eyes' on wings and tail. Females are smaller and browner. Common resident in the western forests, where the male's loud harsh calls, repeated continuously for a minute or two, soon become familiar; also found in the north, where it is rarer.

GREAT ARGUS *Argusianus argus* 76–200cm

The ringing 'kwow-wow' call of this pheasant may be heard at a considerable range, but the bird itself is rarely seen. The male is a large brown bird, with the markedly extended secondaries patterned with 'eyes'. The central tail feathers are greatly elongated, accounting for the enormous length. Females are a rich brown, heavily spotted and streaked. Both sexes have bare blue skin on head and neck. An uncommon resident in the lower hill-forests of the peninsula, commoner in the extreme south.

Male (above); female (below)

GREEN PEAFOWL *Pavo muticus* 102–245cm

The largest and most spectacular of all Thai birds. Male has display of a vast fan-shaped tail with eye-like markings. The male's call, 'toong-hoong' (second syllable stressed), is most often heard at dawn or dusk; also gives a loud braying call. Likes mixed woodland and lowland clearings near rivers. It is now rare, probably restricted to Huai Kha Kheng and Thung Yai Naresuan Wildlife Reserves, where it is resident (a small population recently reported in the north).

SCALY-BREASTED PARTRIDGE *Arborophila chloropus* 29cm

The commonest of a group of medium-sized, plump, short-tailed birds, generally brownish, with striking head patterns. The Scaly-breasted Partridge has no striking wing pattern or belly pattern. The crown and long eye-stripe are dull grey-brown; the equally long superciliary stripe and the throat are white, heavily spotted with black; breast brown with darker vermiculations, and belly orange-rufous. Common resident in most forest areas, except in the peninsula.

SLATY-BREASTED RAIL *Gallirallus striatus* 27cm

Rails tend to be shy skulkers, rarely seen except at dawn or dusk when they may emerge from the reed cover on to open mud to feed. The crown and hindneck of this species are chestnut, the throat and breast grey. Upperparts and flanks are dark grey, barred whitish. Call is a sharp 'cerrk' or 'currk'. This is probably the commonest of the group in Thailand. It is resident in the central plains, where it inhabits swamps and edges of rice paddies, and in coastal marshes and mangroves in the south of the country.

WHITE-BROWED CRAKE *Porzana cinerea* 20cm

This little crake (or rail) can be identified by the black line through the eye, with white lines both above and below it. The rest of the head and the underparts are grey, with buff flanks; the back is brown, streaked blackish. Legs pale green. Inhabits lakes and marshes, especially those with dense platforms of floating vegetation. Resident in suitable habitats from Bung Boraphet (Nakhon Sawan) south into the peninsula. Like the Slaty-breasted Rail, it is most often seen at dawn or dusk.

WHITE-BREASTED WATERHEN *Amaurornis phoenicurus* 33cm

One of the commonest and most easily observed residents of marshes and ponds. Upperparts dark grey, contrasting with striking white forecrown, face, throat and breast; undertail-coverts pale chestnut. Greenish bill and legs. Short cocked tail is often flicked up and down. Very noisy; utters a strange mixture of squawks, bell-like notes and others. Quite often perches on bushes. Very common throughout Thailand (except the north-east), inhabiting edges of marshes, ponds, paddyfields and mangroves.

WATERCOCK *Gallicrex cinerea* 43cm

A large, long-necked bird of dense marshland. The breeding male is blackish, with buff undertail-coverts, and has a red frontal shield and bill and red legs. The female and non-breeding male are light brown, marked with black, with greenish bill (lacking shield) and legs. Much larger than the rather similar Common Moorhen, and not so clumsy as the Purple Swamphen. Resident on marshes and paddies in the central plains, and a wet-season visitor throughout the country in suitable habitats.

GREY-HEADED SWAMPHEN *Porphyrio poliocephalus* 43cm

A large, bluish-purple marshbird with long red legs and red bill, and frontal shield. It is the same size as a Watercock, but blue rather than black, and much heavier and clumsier-looking. White undertail-coverts. Prefers the less disturbed parts of large marshy areas, or dense floating vegetation on larger lakes. Found predominantly in the north and centre of Thailand. Very similar to the smaller Black-backed Swamphen (*P. indicus*), which is more common in the south and does not show silver on head and neck.

MASKED FINFOOT *Heliopais personata* 53cm

A large and unusual waterbird with a long neck and a pointed yellow bill. The male's throat and foreneck are entirely black, bordered each side by a white line. The female has a white patch in the centre of the black. Rest of the upperparts greyish-brown. When swimming, the body is partly submerged (not entirely as the Oriental Darter). This shy species likes slow-moving streams, and also occurs in mangrove forests. Scarce passage migrant and winter visitor in the peninsula.

PHEASANT-TAILED JACANA *Hydrophasianus chirurgus* 30cm

Best known for its ability to walk with ease across floating vegetation, it can be identified in all adult plumages by the golden-yellow hindneck and the black line down the sides of the neck. In flight, the wings are white with a black tip, and the legs are trailed behind. The male in breeding plumage has an extremely long black tail, and a black body with a large white wing patch. Resident and winter visitor in marshy areas and lakes in many parts of the country.

BRONZE-WINGED JACANA *Metopidius indicus* 28cm

This is a darker relative of the Pheasant-tailed Jacana, always with a short tail. The body is black, with bronze wings and back and reddish rear end, and it has a distinctive white supercilium. It lacks the white wing patch of the Pheasant-tailed, and the neck is entirely glossy black. Juveniles are white below, with a strong buff wash to the breast and neck. A common resident of marshes and weed-covered ponds in the central plains area, and also found in suitable areas in the north and south of the country.

PACIFIC GOLDEN PLOVER *Pluvialis fulva* 25cm

The mottled brown upperparts, chunky build and, in breeding plumage, black face, breast and belly distinguish this from most other wading birds. In flight, a slight white wingbar. It may be found in winter on mudflats with other waders, and also on grassy fields inland. The Grey Plover *Pluvialis squatarola*, which is larger and greyer, also has a black belly in summer and keeps more to mudflats and sandy shores; it has striking black 'armpits' in flight.

LITTLE RINGED PLOVER *Charadrius dubius* 18cm

The most widespread of a group of small waders all of which have a complete white collar in all plumages. It has yellow to orange-yellow legs and a yellow orbital ring, and a complete black or brown breast band, incomplete on some immatures. It shows no wingbar in flight. Although it occurs on the shore, it is also found commonly inland near lakes, rivers and ponds. Primarily a winter visitor, but small numbers are resident, the latter breeding mainly along rivers in the north.

KENTISH PLOVER *Charadrius alexandrinus* 15cm

Superficially similar to the Little Ringed Plover, this species can be distinguished at all times by its smaller size, broken breast band, black legs, and white wingbar in flight. Breeding male has a rufous-chestnut cap, with black forecrown and eye-stripe. A very common winter visitor, mostly in coastal areas. The very similar Malaysian Plover *Charadrius peronii* has more mottled upperparts, pinkish-grey legs, and is uncommon and restricted to sandy beaches in the peninsula.

GREATER SAND-PLOVER *Charadrius leschenaultii* 24cm

This and the Lesser Sand-plover *Charadrius mongolus* are the common plovers of the mudflats, brownish in winter and without distinctive markings. In summer, the Greater Sand-plover has more white than the Lesser on a mainly black forehead, and its orange breast-band is duller, usually narrower in the centre and not so clearly demarcated from its white throat. The best distinctions between the two species are bill shape (longer and heavier on the Greater) and length of upper leg (longer on the Greater); the two species often form mixed flocks. Common winter visitor.

WHIMBREL *Numenius phaeopus* 43cm

A large, heavily mottled brown wader with a decurved bill and striped crown. Whitish rump in flight, extending in a V up the back. When it takes flight, the Whimbrel frequently utters a distinctive trilling call. Generally the largest wader on the mudflats, unless the even bigger Eurasian Curlew *Numenius arquata* or Eastern Curlew *Numenius madagascariensis* are present; both these species have much longer decurved bills. Common winter visitor along the coast.

ASIAN DOWITCHER *Limnodromus semipalmatus* 35cm

This species and the two godwits *Limosa* spp. can be told from other large waders by their long, almost straight bills. The dowitcher has a black bill, somewhat thickened at the tip, and normally held at an angle downwards; it is chunkier and shorter-necked than the godwits, which have a flesh-coloured base to the bill. Whitish rump in flight. It feeds by a so-called 'stitching' action. One of the rarest of Palearctic waders; an uncommon passage migrant to the coastal mudflats. Found in mixed flocks of waders, usually in shallow water, not on mud.

COMMON REDSHANK *Tringa totanus* 28cm

The sentinel of the mudflats, usually the first species to take alarm. This is the common red-legged wader of the coastal mudflats in winter, often seen singly or in twos and threes along muddy creeks. Brown above, whitish below, with a long straight bill. The loud ringing call, 'teuuu, teu-heu-heu', soon becomes familiar. In flight, it shows a dramatic white bar along the rear of the wing, and a white rump. Very common winter visitor to coasts, sometimes visiting inland marshes and flooded areas.

COMMON GREENSHANK *Tringa nebularia* 35cm

The grey upperparts and white underparts, as well as the more elegant appearance, distinguish this long-legged wader from the Common Redshank. The legs are greenish. The bill is very slightly upcurved. No wingbar, but a V-shaped wedge of white up the back in flight. Ringing call all on one note, 'teu-teu-teu', usually given in flight. Common winter visitor on the coastal mudflats and also on freshwater marshes; tends to feed in shallow water rather than on the bare mudflats.

MARSH SANDPIPER *Tringa stagnatilis* 25cm

Similar to a small, slender and hyperactive Common Greenshank, but with a noticeably slender, straight black bill. Marsh Sandpipers feed in the shallows, often scampering to and fro; in flight, the rapid piping call is distinctive. The flight pattern is similar to that of the Common Greenshank, with conspicuous white wedge extending up the back and contrasting with dark wings, but the legs project well beyond the tail. Marsh Sandpiper is a common winter visitor to the coast and freshwater marshes.

GREEN SANDPIPER *Tringa ochropus* 24cm

The striking black and white appearance of this bird in flight is distinctive: white rump, but no white V up the back; the dark underside of the wing separates it from the Wood Sandpiper. At rest, it is darker and less spotted on the upperparts; legs greenish. Has intermittent tail-bobbing action. Gives characteristic call, 'klu-eet-weet-weet'. Rather common winter visitor, found on the edges of ponds, lakes and rivers and in ditches; generally solitary, though several may occur in a small area.

WOOD SANDPIPER *Tringa glareola* 23cm

The common sandpiper of the freshwater marshes, often occurring in flocks. At rest, it is paler and more heavily spotted than the Green Sandpiper, and has a fairly prominent long white supercilium; in flight, the underside of the wing is pale and the tail is more strongly barred, so the white rump, although still obvious, appears smaller. Legs yellowish. The call is a rapid 'chiff-chiff-chiff', usually given on take-off and in flight. Very common and in places abundant winter visitor on inland marshes and paddyfields, as well as coastal mangroves and ponds.

COMMON SANDPIPER *Actitis hypoleucos* 20cm

This species has a characteristic flight comprising rapid, stiff wingbeats interspersed with short glides as it flits along a river bank or across a pond. Rather uniform olive-brown above and white below, with indistinct eye-ring and supercilium and dark patches on sides of breast. Prominent white wingbar in flight. On ground, repeatedly bobs tail. High-pitched 'wee-wee-wee' call given in flight. Normally seen singly or in twos and threes along the edges of rivers, ponds, or at the edge of the mudflats. Very common winter visitor throughout the country.

RUDDY TURNSTONE *Arenaria interpres* 23cm

In breeding plumage, the reddish-brown back, boldly marked with black, and the black and white head pattern are unmistakable. In non-breeding plumage, it is dull brown, heavily streaked, with a dark breast. Legs relatively short, and bright orange. Dumpy silhouette is very different from that of the smaller stints and sandpipers with which it associates. In flight, it has a double white wingbar, as well as white on the lower back and at the base of the tail. Common winter visitor on the coast.

RED-NECKED STINT *Calidris ruficollis* 16cm

Scampering flocks of several hundred small waders on the seashore in winter usually consist mainly of this species. Single birds can be identified by the short bill, dark legs and rather scaly-looking pale greyish-brown upperparts. Similar species include Temminck's Stint *Calidris temminckii*, more uniform and darker grey above, legs yellowish, prefers fresh water; and Long-toed Stint *Calidris subminuta*, more heavily mottled above, legs yellow, shape more sandpiper-like than stint-like, prefers grassy edges of freshwater marsh. All have thin white wingbars, and are common winter visitors.

CURLEW SANDPIPER *Calidris ferruginea* 22cm

The commonest of a group of waders slightly larger than the stints. The best distinguishing features are the slightly decurved bill and, in flight, the white rump patch and absence of wingbar. In winter, grey above and white below, with fairly prominent white supercilium. In breeding plumage the underparts become deep chestnut. Frequently in large flocks, or forming the majority of individuals in mixed flocks of small waders. Usually prefers coastal mudflats, where it is a common winter visitor.

PINTAIL SNIPE *Gallinago stenura* 25cm

With their camouflaged plumage and unobtrusive habits, snipe are not easily seen on the ground. All species of snipe have long, straight bills and heavily streaked brown plumage. The flight of the Pintail Snipe is heavier and more direct than that of the Common Snipe, and it also has a shorter and weaker call than the latter. This species also prefers drier ground than the Common Snipe, and may be found high up in the hills as well as in the lowlands. Common winter visitor.

COMMON SNIPE *Gallinago gallinago* 28cm

This is the familiar snipe species of Europe, known for its zigzagging flight when flushed. In Thailand it is the common snipe of marshes and paddyfields, and in winter may be seen anywhere in the country where there is suitable habitat. The harsh and rasping call, 'scaap', is almost invariably uttered when flushed, when the bird flies away rapidly on a noticeably erratic course with sudden lunges from one side to the other, often rising high in the air. Very common winter visitor.

BLACK-WINGED STILT *Himantopus himantopus* 38cm

A striking black and white bird with a long, black, needle-like bill and immensely long pink legs, which trail behind it in flight. It is usually found in small parties on freshwater marshes or open ponds and lakes, sometimes on the shore. Has a sharp, high-pitched 'keek' call, which is frequently repeated when the bird is nervous or alarmed. A local resident in the south and a fairly common winter visitor in most suitable habitats throughout the country.

LITTLE TERN *Sterna albifrons* 23cm

The smallest of all the terns in Thailand. Identified by its black-tipped yellow bill and yellow legs in breeding plumage, its small size, and its habit of hovering with very fast wingbeats, and then plunging almost vertically into the water. In non-breeding plumages the bill is entirely black and the legs much duller in colour. It has a white forehead in all plumages. Often noisy, uttering short chattering 'kik-ik' notes, often in repeated series. A common resident of coastal areas in the south of the country.

WHITE-WINGED TERN *Chlidonias leucopterus* 23cm

This and the Whiskered Tern (*Chlidonias hybridus*) are the common passage and wintering terns along the southern coasts. Outside the breeding season the two are very similar, and their plumages extremely confusing, especially as birds seen in Thailand are often in immature or intermediate plumages. In most plumages the hindcrown of White-winged Tern is blackish (white speckled with black on Whiskered), and the rump is whitish (not grey). Whiskered is much more likely to be seen away from the coast; whereas the White-winged has a preference for coastal waters.

THICK-BILLED GREEN-PIGEON *Treron curvirostra* 27cm

A large green pigeon with a distinctive 'swollen' bill. Like many of the green pigeons which make up the genus *Treron*, the male has a maroon mantle and yellowish wingbars. Identified at close range by its heavy bill, red at the base and green at the tip, and its blue-green orbital ring. A common resident of inland forests, giving way to Wedge-tailed Green-pigeon *Treron sphenura* at higher altitudes. It is largely absent from the central plains and the north-east.

LITTLE GREEN-PIGEON *Treron olax* 20cm

The smallest and darkest of the green pigeons. The male has a dark grey head and an orange breast band; the back is a dark maroon, the undertail-coverts dark chestnut, and the tail is dark grey with a paler grey tip. The female is dark green above, dull green below. Utters soft whistles like other green pigeons, but this species' voice is higher-pitched. Rare resident of the evergreen forests of the lowlands; restricted to the peninsula.

JAMBU FRUIT DOVE *Ptilonopus jambu* 27cm

The male of this attractively patterned green pigeon has white underparts, with a pink patch on the breast. The face and crown are crimson, duller and less distinct on the female, which is otherwise all green, with white only on the centre of the underparts. Both sexes have an orange bill and a prominent white ring around the eye. An uncommon resident of forests and mangrove swamps, in Thailand seen only in the extreme south where it is threatened by habitat destruction.

GREEN IMPERIAL PIGEON *Ducula aenea* 43cm

This large and sturdy bronze-green pigeon has a grey head and underparts, with a chestnut vent. The latter feature and also the uniformly dark tail distinguish it from the otherwise very similar Mountain Imperial Pigeon, which prefers higher elevations but may occasionally descend to lower levels. The Green Imperial Pigeon is a resident of lowland forest, and survives only in a few areas scattered throughout the country, where it may, however, be numerous.

MOUNTAIN IMPERIAL PIGEON *Ducula badia* 47cm

This is much the largest pigeon of the northern and western mountains. The upperparts and wings are deep brown (with maroon tinge), the head and underparts grey. The vent is pale creamy-buff, and the dark tail has a broad pale terminal band which is most noticeable in flight. Imperial pigeons often announce their presence by their deep booming calls. Common resident of evergreen forests on the mountains of the north, west and south-east; occasionally visits nearby lowland areas.

ROCK PIGEON *Columba livia* 33cm

The Rock Pigeon is the common feral pigeon found throughout most of Europe and Asia. The plumage varies considerably; the original type is dark grey with black bars on the wings and a black terminal band on the tail. Although Rock Pigeons are commonest in urban and cultivated areas throughout the country, they can still be found on the rocky cliffs of the higher mountains, where the original (introduced) stock may sometimes be purer. Resident.

LITTLE CUCKOO-DOVE *Macropygia ruficeps* 30cm

With their long tails, the cuckoo-doves look rather like cuckoos, especially in flight. The crown is orange-rufous, duller on the female. The upperparts are dark brown, heavily scaled with rufous; the tail is unbarred. The underparts are reddish-buff, the breast with paler scaly mar[illegible]gs (male) or with black mottling (female). The underwing- [illegible] right cinnamon-orange. An uncommon resident in the [illegible] northern mountains, above 500m.

RED COLLARED-DOVE *Streptopelia tranquebarica* 23cm

The small size and reddish plumage differentiate this species from its relatives. It has a narrow black bar across the base of the hindneck, and plain (not scaly-looking) upperparts. The male is reddish, with blue-grey head; the female dull brown, with head duller grey. In flight, both sexes show a comparatively short dark tail with an incomplete white band at the tip (broken in the centre). A common resident in open country, except in the peninsula.

SPOTTED DOVE *Streptopelia chinensis* 30cm

The commonest of all open-country pigeons, this species can be found even in the towns. The black neck patch (on hindneck) is heavily spotted, not barred, and the whole bird is much paler than the Oriental Turtle-dove *Streptopelia orientalis*, which has white bars on the neck patch (on side of neck) and noticeably scaly-patterned upperparts. The tail is wedge-shaped, with the tips of the outer feathers broadly white (white-cornered tail), a clear field mark. Very common resident everywhere.

ZEBRA DOVE *Geopelia striata* 21cm

As this is a common cagebird, it may occur anywhere. It is very small, much smaller and slimmer than the Red Collared-dove, and heavily barred on the neck and down the flanks; the brownish upperparts are also barred, but less obviously. The head is grey, with brown hindcrown and nape, and with a broad bluish orbital ring. Frequently perches on telegraph wires. Common resident in the peninsula and in the central plains.

EMERALD DOVE *Chalcophaps indica* 25cm

The metallic glossy green wings and back are the diagnostic features as the bird flies off through the undergrowth; the reddish-chestnut wing-linings may also be visible. This is characteristically a bird of the forest undergrowth, seen mostly in flight, or feeding in the litter of the forest paths. Its call, a quiet, plaintive 'hooo' repeated at short intervals, is a commonly heard sound in areas where this bird occurs. Common resident in most forested areas up to about 1500m.

NICOBAR PIGEON *Caloenas nicobarica* 41cm

The greenish-bronze plumage of this large ground-loving pigeon is familiar to many people from bird collections throughout the world. It is a plump-looking bird with a short white tail and red legs. The adult has a ruff of glossy bronze-green hackles, lacking on the immature. Having been much hunted over many years, this species is now rare in the wild, being confined to the least frequented islands off the west coast, where it is resident in evergreen forests.

BLUE-CROWNED HANGING PARROT *Loriculus galgulus* 14cm

This tiny parrot often hangs upside-down as it forages for food. It is green, with a black bill, red rump, and greyish feet. The male has a yellow patch on the mantle and another above the rump, and a red patch on the breast. The male has a small blue patch on the crown. The tail is short. This is a common resident in forested areas in the extreme south. In much of the rest of the country it is replaced by the Vernal Hanging Parrot *Loriculus vernalis*, which resembles the female Blue-crowned but has a red bill and lacks the blue crown patch.

LARGE HAWK-CUCKOO *Hierococcyx sparverioides* 44cm

Where this species breeds, its incessant call, 'pipeeah' (or 'brain fe-ver'), is very familiar. The bird itself is rarely seen. The adult is slate-grey, with dark-streaked rufous breast and dark-barred whitish belly. Boldly banded tail. Young birds are brown above, heavily barred, and white with heavy streaking below. In flight, rather like a sparrowhawk, but bill longer and slightly decurved. Resident in forests in the north; winter visitor in central plains, east, and parts of the peninsula.

ASIAN KOEL *Eudynamys scolopaceus* 43cm

The loud 'koel' call which gives this bird its name is a familiar sound in villages and towns throughout the country except in the north-east and far north. The male is a large, long-tailed, entirely glossy black bird with a red eye; the female, is also long-tailed and red-eyed, has a dark brown plumage and is heavily marked with buff and whitish bars and spots. The male often perches in exposed locations, while the female is much harder to see. Common resident around human habitations, in open woodland, gardens and scrub.

Male (above); female (below)

GREEN-BILLED MALKOHA *Rhopodytes tristis* 56cm

This is a large, clumsy, greyish bird with an inconveniently long tail. It is dark grey above, paler below, with a large red eye patch and white tips to the tail feathers. The bill is greenish. It may be seen in the upper branches of the trees; characteristically, however, it enters a dense clump of foliage near the bottom and works its way up inside, before emerging at the top. Common resident in forest and scrub areas throughout the country, except on the highest mountains of the north.

RAFFLES' MALKOHA *Rhinortha chlorophaeus* 33cm

The smallest malkoha, and the only one which is sexually dimorphic. The male is chestnut above, pale rufous below, and has a blackish tail with white tips to the feathers. The female has a grey head and breast, and the tail is chestnut (like the back) with black and white tips to each feather. Locally common resident in lowland forest and the forest edge up to 900m in the peninsula, and at Kaeng Krachan and Thung Yai Naresuan.

GREATER COUCAL *Centropus sinensis* 53cm

A large, long-tailed, crow-like, black bird with chestnut wings and back which may be seen commonly almost anywhere in scrub, in grassland, and near habitations. It often walks on the ground, and perches freely on clumps of bamboo and even on rooftops. The call, a series of hooting 'poop poop poop poop...' notes descending in pitch and sometimes rising again at the end, is one of the common bird sounds of rural Thailand. Very common resident.

LESSER COUCAL *Centropus bengalensis* 38cm

This species is much smaller than the Greater Coucal and rather less brightly coloured, but with underwing-coverts chestnut (black on Greater). In the non- breeding and juvenile plumages it is browner above, heavily streaked and barred, and much paler below. The call is distinctive, a series of 'whoop' notes followed by a short series of higher-pitched, ringing 'kotok kotok' notes. A common resident throughout the country, but it is found near houses less often than Greater Coucal; it also prefers marshy land.

BARN OWL *Tyto alba* 34cm

This is the 'white owl' of caves and farm buildings. It is light golden-brown above, with only small, indistinct markings, and white or whitish below, and has a very distinctive heart-shaped white facial disc and dark eyes. One of the easiest owls to see, as it comes out to hunt shortly before dusk. Has a variety of calls, including a loud rasping or hissing sound. Common resident throughout the country, most often hunting over open areas and marshes. Roosts in caves on the hillsides.

BROWN BOOBOOK *Ninox scutulata* 30cm

The tail, longer than that of most owls, gives this species a distinctive silhouette. It is also the only medium-sized owl without a facial disc. Dark brown above; pale below, with broad brown streaks. Call a double whistle, 'hoo-op', with the second note higher in pitch, repeated at intervals of a second or two. Occurs mostly in forests and wooded areas, but also in more open areas with some trees. Common resident and winter visitor throughout the country with the exception of the north-east.

SPOTTED WOOD-OWL *Strix seloputo* 48cm

More often heard than seen, this large nocturnal species utters a deep, powerful and explosive 'who' (D.R. Wells). Wood-owls have no ear-tufts (unlike most other owls). This species, if seen, is best identified by the white spotting on the upperparts, and heavy dark barring on the underparts; the crown is also spotted. The facial disc in contrast is plain pale rufous. Prefers open woodland, secondary growth and mangroves. Uncommon resident, most likely to be found in the peninsula.

LARGE-TAILED NIGHTJAR *Caprimulgus macrurus* 30cm

Nightjars are nocturnal birds, and therefore much more frequently heard than seen. The call of this species is a loud and resonant 'chonk', uttered at one-second intervals while the bird is at rest. In flight, the male is recognizable by the distinct white wing patches and the prominent white corners to the tail; on the female, these markings are much duller (more buff) and far less obvious. Common resident in open country, up to 2000m, throughout Thailand except near Bangkok.

SCARLET-RUMPED TROGON *Harpactes duvaucelii* 25cm

Trogons are medium-sized, shy birds of the forests; most species are orange above and red below, the males with black heads. The best field marks of this rather small species are the bare pale blue skin above and in front of the eye and the scarlet rump (duller on the female). The female's head is dark brown. The territorial call is a series of a dozen or so short 'tewk' notes, speeding up and dropping in pitch. An uncommon resident of lowland forests in the southern half of the peninsula.

RED-HEADED TROGON *Harpactes erythrocephalus* 34cm

The red head of the male and the white crescent across the breast on both sexes make this trogon easy to identify. In two other trogons, the Red-naped *Harpactes kasumba* and Diard's *Harpactes diardii*, the male also has a pale or white crescent across the breast, but both have a black head and upper breast, and they occur only in the far south. The Red-headed is quite common in the hill-forests of the north, west and south-east, but is not found in most of the peninsula.

COMMON KINGFISHER *Alcedo atthis* 17cm

The brilliant turquoise back and rump as this kingfisher flies across the water are unmistakable. This is the familiar species of Europe and Asia. It is the smallest of the commoner kingfishers. Note the black bill (often with red at base of lower mandible), the dull orange underparts and, at close range, the rufous ear-coverts. Frequently seen diving for fish from a branch of a bush overhanging the water. Very common winter visitor throughout the country; a very few are resident.

STORK-BILLED KINGFISHER *Pelargopsis capensis* 37cm

A spectacular blue-green kingfisher with dull orange underparts and collar, a mostly brown head, and a massive red bill. It frequents the banks of forest rivers, and also lakes, preferably with scattered trees. Likes to perch on low branches over the water. The White-throated and Black-capped Kingfishers are smaller, and in flight reveal a white wing patch, which the Stork-billed lacks. Uncommon resident although can be locally fairly common in most parts of the country except the north-east.

RUDDY KINGFISHER *Halcyon coromanda* 25cm

One of very few kingfishers which has no blue in its plumage, though it may show a hint of blue on the pale-coloured rump. The Ruddy Kingfisher has reddish-violet upperparts, including head and tail, and bright rufous underparts. The bill is a striking bright red. The whitish uppertail-coverts and lower rump are revealed in flight. A bird of the islands, mangroves and lowland forest streams. Uncommon resident in the peninsula; winter visitor and passage migrant elsewhere.

WHITE-THROATED KINGFISHER *Halcyon smyrnensis* 28cm

The brilliant turquoise-blue of this bird is a familiar sight throughout the country, sometimes quite far from the nearest water. The throat and breast are white and the head and belly chestnut-brown. In flight, it shows a conspicuous white wing patch and a dark wing-covert band. Like the Black-capped and Collared this one does not dive for fish; it feeds on insects, small frogs etc. A loud, ringing, staccato call, like a shrill laugh descending in pitch, often betrays its presence. Very common resident in open country and open woodland.

BLACK-CAPPED KINGFISHER *Halcyon pileata* 30cm

The deep violet-blue of its back and wings makes this the most spectacular of all the kingfishers. It has a black head and a narrow white collar; the throat is white, shading into bright pale rufous on the belly. The bill is bright red. Like all *Halcyon* kingfishers, this species likes to perch in fairly exposed places; it is often seen perched by a stream or lake. Frequents paddyfields and swamps, as well as mangroves. Common passage migrant and winter visitor throughout the country, mainly in lowlands.

COLLARED KINGFISHER *Todiramphus chloris* 24cm

A small kingfisher, greenish-blue above and white below, with a conspicuous white collar. The bill is black with a paler, flesh-coloured lower mandible. Lacks wing patches in flight. The territorial call is a raucous staccato laugh on a slightly descending scale, terminating in highly distinctive 'jee-jaw' notes. This species is a very common resident along all the coasts, being a characteristic bird of mangroves and of the casuarina windbreaks above the tideline; occasionally seen further inland.

BLUE-TAILED BEE-EATER *Merops philippinus* 30cm

An attractive species, mainly green, with a bold whitish streak below the black eye-line; lower throat coppery-brown. Elongated central tail feathers. In flight, reveals blue rump, uppertail-coverts and tail. Inhabits open country, with strong preference for vicinity of water. Nests in colonies, usually in river banks. It may be found in most parts of the country except the north-east: in summer in the north, year round in the central areas, and on passage in the peninsula.

LITTLE GREEN BEE-EATER *Merops orientalis* 20cm

This elegant little bee-eater is by far the commonest member of its family throughout the north. Mainly green, with a brown crown and nape and a black gorget. Elongated central tail feathers. Nests in burrows, often dug in flatter ground. Generally seen in small groups in scrubland or near ponds in dry open country, mostly in lowlands, but ranges up to 1500m. Often visits parks and gardens. Very common resident throughout the country except the central plains and peninsula.

BLUE-THROATED BEE-EATER *Merops viridis* 28cm

The darkest and bluest of the *Merops* bee-eaters. This species has a blue throat with no gorget, and a deep chestnut-brown crown, nape and upper back. Elongated central tail feathers. In flight, it shows light blue uppertail-coverts and rump and darker blue tail. Juveniles have the crown and upper back green, not brown. Found mostly in lowlands, in open forested areas, scrub and mangroves. Fairly common passage migrant and winter visitor to the peninsula and the south-east.

INDIAN ROLLER *Coracias benghalensis* 33cm

A characteristic bird of roadside telegraph wires and other exposed perches, from which it drops to the ground to catch insects, lizards, etc. Sometimes also catches insects in flight. At rest it looks mainly chestnut brown, with patches of turquoise. In flight, however, the wings and tail show much brilliant dark blue and pale blue, with a light blue rump. Gives a crow-like croaking call. A common resident of open dry areas throughout the country, up to about 1500m.

DOLLARBIRD *Eurystomus orientalis* 30cm

The characteristic flattened head of this species makes it identifiable from a considerable distance. Plumage appears all blackish, but with a pale silvery-blue patch on the outer wings (said to recall a silver dollar) in flight. Red bill. It feeds largely by catching insects in flight. Like the Indian Roller, it likes exposed perches such as telegraph wires. A common resident and winter visitor in deciduous woods and forest edge in most of the country; absent from most of the north-east and the central plains.

HOOPOE *Upupa epops* 30cm

No other bird possesses quite the pinkish-brown colour of the Hoopoe's head, shoulders, crest and breast. The long crest is only seldom raised into a fan. The wings are barred black and white, this and the white band on the black tail being particularly striking in flight (when wings also look very broad and rounded). The bill is long and decurved. Feeds on the ground, rarely perching on trees or bushes. The name is a representation of its call. Common resident in open country and cultivated areas.

ORIENTAL PIED HORNBILL *Anthracoceros albirostris* 70cm

Although the smallest of the hornbills, this is still a very big bird. It has the large bill and casque characteristic of the family. Its plumage is mainly black, with a white belly, white tips to the outer tail feathers and a white trailing edge to the wings in flight. Usually seen in small groups in mostly evergreen forest. Now restricted mainly to the larger blocks of forest down the western boundary of the country, with a few elsewhere (notably Khao Yai and Khao Sam Roi Yot). Resident.

GREAT HORNBILL *Buceros bicornis* 122cm

This enormous bird announces its approach from afar by its loud calls and the whoosh of its wingbeats. The huge bill and casque are yellow, the face is black, and the neck is a washed-out golden-yellow. In flight, the yellowish bar across the full length of the wings and the broad white trailing edge are easy to recognize. The tail is white, with a single broad black band across the centre. Found mostly in the same areas as the Oriental Pied Hornbill. Locally common resident.

GREAT BARBET *Megalaima virens* 32cm

This largest of the barbets may be heard throughout the mountain forests of the north: its call is a ringing 'kayoh-kayoh-kayoh', which can be heard over a considerable distance. Calling birds perch near the tops of forest trees, but are often difficult to locate, as they are accomplished ventriloquists! The bird is mainly dark greenish and brown, with belly and flanks green-streaked yellowish and vent red, but the most obvious feature is its massive yellow bill. Common resident in forests above 600m.

LINEATED BARBET *Megalaima lineata* 29cm

This species' pale yellowish-straw head and upper breast, streaked with brown, are distinctive. The bill is yellow; dark eye, encircled by broad orange-yellow orbital ring, is conspicuous. The call is a melodious 'poo-poh', the second note delivered at a higher pitch. In most of the country this is one of the commoner lowland barbets, preferring the more open forest and cultivated areas. Resident throughout the country, being absent from most of the north-east.

RED-THROATED BARBET *Megalaima mystacophanos* 23cm

Most barbets are green, with complex and colourful head patterns; this species is the only one in which the male has a red chin and throat. The forehead is yellow, the hindcrown red. Very heavy black bill. Barbets in general are chunky birds of the treetops, usually recognized by their loud, monotonous calls. Red-throated utters a slow series of deep single and double notes in alternating sequence, 'chok, chok-chok, chok,...'. Common resident of forests in the peninsula.

BLUE-THROATED BARBET *Megalaima asiatica* 23cm

This species shows much more blue on the throat and sides of the head than any other barbet. The forecrown and hindcrown are red, separated by a narrow black central band which also extends down the sides of the nape. The bill is black, with a paler base. It is probably the most common barbet of the northern evergreen forests, where its 'took-arook' call is uttered incessantly. Not normally found below 600m or above about 1600m. Common resident in the north and west.

COPPERSMITH BARBET *Megalaima haemacephala* 16cm

The incessant 'tonk-tonk-tonk' call note of this species is a frequently heard sound of the lowlands, including towns and gardens. This is the smallest barbet and much the easiest to see. Red forecrown, yellow eye patch bordered below and behind with black, and yellow throat and upper breast divided horizontally by a red band. The rest of the underparts are yellowish, broadly streaked dark green, and the upperparts plain dark green. Very common resident throughout the country except in dense forest.

SPECKLED PICULET *Picumnus innominatus* 10cm

This tiny olive-green bird is no bigger than a flowerpecker, and has head striped black and white and underparts which are entirely barred black and white. It is a frequent component of feeding flocks (bird waves). Difficult to see, but it makes its presence known by a frequent loud tapping noise as it searches for food. The upperparts are olive-green, and the extremely short tail is black and white. Common resident of the northern evergreen and mixed forests, being found up to 1800m.

COMMON FLAMEBACK *Dinopium javanense* 30cm

Male

Female

This spectacular woodpecker is a riot of colours: crown red (black with dense white streaky spots on the female), head striped black and white, underparts scalloped black and white, wings golden-brown and back red. Distinguished from the similar Greater Flameback *Chrysocolaptes lucidus* by the much shorter bill, and by the single (not double) moustachial stripe. It prefers open deciduous woodland, and is resident throughout most of the country, though absent from much of the north-east.

LACED WOODPECKER *Picus vittatus* 30cm

One of a group of closely related species which are green above, and usually with scaly markings on the underparts. Crown red (male) or black (female). This species has an unmarked yellowish upper breast, with scaly markings covering the rest of the underparts. Found in every kind of wooded country, from mangroves up to the evergreen forests of the mountains (to 1500m). Common resident in most of the country; absent from the north-east (except along the Mekong) and the peninsula.

GREATER YELLOWNAPE *Chrysophlegma flavinucha* 34cm

A large green woodpecker with a bright orange crest and yellow under the chin, the yellow being less extensive on the female. The head is otherwise dark, with dark chestnut crown. The upperparts are plain greenish-olive, and the tail is black; the primaries are a dark rufous-chestnut colour, barred black. The underparts are dull grey, with no barring. This is a common resident of forested areas throughout the country except the peninsula, the far north, and the central plains.

CRIMSON-WINGED WOODPECKER *Picus puniceus* 25cm

One of a spectacular group of woodpeckers with crimson wings. This species has a bright red crown and yellow crest, green sides of head and back, and crimson wings. The underparts are green and unmarked, except for some white barring on the lower belly. The male has a short red moustache. Drums weakly. Restricted to the southern half of the peninsula, where it is a common resident of lowland forest and secondary growth and forest plantations up to 600m.

MAROON WOODPECKER *Blythipicus rubiginosus* 23cm

This species also has reddish wings, though far less bright than those of the Crimson-winged Woodpecker, but it has no crest. Generally when viewed it appears dark. The head is brown, the male having a red patch on the side of the nape and neck; the underparts are dark brown. The bill is pale yellowish, contrasting with the dark plumage. Does not drum. The distribution is similar to that of the Crimson-winged, but it is normally confined to evergreen forest, and extends up to about 900m.

GREAT SLATY WOODPECKER *Mulleripicus pulverulentus* 50cm

This large slaty-grey bird is the biggest of the Old World woodpeckers. Unmistakable because of its large size and uniformly grey plumage; the throat is a contrasting orange-yellow. The male has a small red moustache. Does not drum, but instead has a loud braying call (like a donkey). It is characteristically found in pairs, or small family groups, in rather open lowland forest. Uncommon to scarce or rare resident throughout the country except the north-east and central plains.

WHITE-BELLIED WOODPECKER *Dryocopus javensis* 43cm

Almost as big as the Great Slaty, this is the most glamorous of all Thailand's woodpeckers. Black, with a white belly and white lower back and rump (upperparts entirely black in peninsular population). Crown and crest red on the male, which also has a red moustache; the female has only hindcrown and crest red. Prefers more open parts of lowland forest, up to about 600m. Although not common, it is fairly conspicuous when present; e.g. Nam Nao. Resident in lowland forests throughout the country.

SILVER-BREASTED BROADBILL *Serilophus lunatus* 18cm

The silvery-grey underparts make this chunky little bird easy to recognize in the middle storey of the forest. The upperparts are mainly rufous-chestnut; it has a broad black eye-stripe contrasting with the pale head, and black and pale blue markings on the wings. The tail is black, with white-tipped outer feathers. Yellow orbital ring. This is a fairly common resident in evergreen forest between 300m and 1800m through most of the country, but is absent from the north-east and from much of the peninsula.

LONG-TAILED BROADBILL *Psarisomus dalhousiae* 28cm

This bird's bright green body and blue tail attract immediate attention. The head is black above, orange-yellow below, with a strong yellowish bill. The yellow throat is sharply demarcated from the green underparts. More commonly seen in flight in the open than other broadbills, when it shows a white patch at base of primaries. Sometimes forms flocks of 40 or 50 birds after the breeding season. Common resident in the forests of the north and south-east as well as in Khao Yai.

BLUE-WINGED PITTA *Pitta moluccensis* 20cm

Pittas are ground-haunting, short-tailed forest birds, usually brightly coloured but unobtrusive, and very difficult to see. This species is green above, with deep blue wings and tail; the head is black, with pale buff stripes each side of the crown, and white chin. The rest of the underparts are deep buff, with red vent. White wing patches in flight. The Blue-winged is probably the commonest and most widespread pitta in the country. Rainy season visitor and passage migrant, with a very few resident.

HOODED PITTA *Pitta sordida* 19cm

A largely green bird, with the head appearing black (the crown is in fact dark brown). The rump and a patch on the wing, visible when the bird is at rest, are turquoise-blue. Lower belly and undertail-coverts red. The sexes are alike. In flight, shows a large white patch on the wing. Prefers moist evergreen forest in the lowlands, occasionally venturing into mixed forest. An uncommon rainy season visitor to the west and south-east of the country; resident and winter visitor in the peninsula.

BANDED PITTA *Pitta guajana* 23cm

The broad bright yellow supercilium, becoming orange on the nape, and the yellowish-white throat are diagnostic of both sexes. Black facial mask and central crown-stripe. The back is bright golden-brown, and there is a very prominent white stripe on the folded wing. The underparts of the male are blue-black, barred with orange on the sides of the breast; the female is barred black and yellow from breast to vent. Uncommon resident in the peninsula, in evergreen forest up to 600m.

HOUSE SWIFT *Apus nipalensis* 15cm

This is the common swift of the towns and cities, often nesting under ledges of buildings. Blackish above and below, with white rump and throat. It is one of two white-rumped swifts; the other, the Fork-tailed Swift *Apus pacificus*, is much larger, with a much bigger wingspan, and is mainly a winter visitor in most of the country (absent from north-east), breeding only in far north. The House Swift feeds mostly over more open areas, and is a common resident throughout the country except the north-east.

BARN SWALLOW *Hirundo rustica* 15cm

The Barn Swallow is best known in Thailand for the enormous winter roosts it forms on the telegraph wires of some major towns. It can be identified in flight by the uniformly dark crown and upperparts and deeply forked tail. Frequently perches on wires, when the chestnut-red forehead, chin and throat and the blue-black breast band can be seen; the rest of the underparts are whitish. A few breed in the northern mountains, but it is common everywhere as a winter visitor.

HOUSE SWALLOW *Hirundo tahitica* 14cm

This is the breeding swallow of the southern coasts. It can be distinguished from the Barn Swallow by the absence of the blackish breast band, the (usually) duskier flanks, and the dark undertail-coverts with white scale-like markings. The underwings are a uniformly greyish colour, darker than Barn Swallow's, and the tail is less deeply forked. Resident in the peninsula and the extreme south-east, where it is found on sea coasts and around villages and towns, and also in open country.

RICHARD'S PIPIT *Anthus richardi* 16–20cm

Richard's Pipit is paler and noticeably longer-legged than other pipits and has a much more upright stance, especially when alert. It is a typical bird of the open fields, grasslands and the margins of paddyfields. Like all pipits, it is brown above, streaked darker (except on rump and uppertail-coverts); there is more or less streaking on the breast, depending on race. Very common resident and winter visitor throughout the country except on the highest peaks. The migrant race is much larger than other pipits.

RED-THROATED PIPIT *Anthus cervinus* 16cm

The characteristic pipit of dry paddies and open areas, usually in small parties. In winter, it is heavily streaked above, including on the rump and uppertail-coverts, and on the breast, and shows a pale V-mark on the mantle. In summer the throat and breast are dull red and unstreaked, though females usually show some streaks (and some may lack red). Stance almost horizontal. Common winter visitor wherever there is open country.

WHITE WAGTAIL *Motacilla alba* 19cm

The common black and white wagtail of Europe and Asia. In winter, it may be seen anywhere where there is open country, even in urban parks. At least three different races are involved, differing chiefly in the amount of white on the face but also in the darkness of the upperparts. Like most wagtails, this species wags its tail up and down constantly, and the flight is noticeably undulating. Found in all kinds of open areas, parks and along rivers. Common winter visitor.

GREY WAGTAIL *Motacilla cinerea* 19cm

This species prefers the vicinity of running water, and in winter may be found near streams throughout the country. The very long tail, grey back, and yellow uppertail-coverts, as well as the striking yellow vent, distinguish it from the Yellow Wagtail; in summer plumage, it has a black throat and often entire underparts bright lemon-yellow. White wingbar in flight. Occurs singly or in pairs, not normally in flocks. Very common winter visitor, usually the first to arrive (in July).

EASTERN YELLOW WAGTAIL *Motacilla tschutschensis* 18cm

Large flocks of this species may be found in marshy areas in the lowlands. In winter plumage it shows very little yellow, less than the Grey Wagtail. The black or dark ear-coverts distinguish it from the Citrine Wagtail in all plumages. In summer, the underparts are bright yellow. No wingbar in flight. Shorter tail than Grey Wagtail. Very common winter visitor to both wet and dry open areas in most of the country, though less numerous in the north.

CITRINE WAGTAIL *Motacilla citreola* 19cm

Female

Breeding male

The brilliant yellow head of the breeding male is diagnostic. Females and non-breeding males have a yellow forehead, and yellowish centres to the ear-coverts; the yellow supercilium extends around the back of the ear-coverts. Juveniles, which lack yellow, have whitish centres to the ear-coverts and a dark patch on the side of the breast. Broad white wingbars, conspicuous in flight. Prefers wetter areas than the Eastern Yellow Wagtail. Fairly common winter visitor to the extreme north and the central plains.

BAR-WINGED FLYCATCHER-SHRIKE *Hemipus picatus* 15cm

This small pied bird of the forests can be confused only with the male Little Pied Flycatcher. Unlike the latter, this species has no white supercilium. It has a completely black crown, whitish throat, and black or dark brown upperparts with white rump and a white wing patch. The underparts are whitish, tinged pinkish. Very active, and often seen in bird waves. A very common resident of primary and secondary growth throughout most of the country, ranging up to all but the highest peaks.

LARGE CUCKOO-SHRIKE *Coracina macei* 30cm

Cuckoo-shrikes are medium-sized black and grey birds frequenting the middle storey of the forest. This is the largest common species. The upperparts are grey, with blackish primaries; it has a rather obscure blackish face-mask. The underparts are also grey, becoming whitish on the lower belly and vent. When it lands, it habitually flicks up each wing alternately. A common resident of the forests, mainly in more open parts, but absent from the peninsula.

GREY-CHINNED MINIVET *Pericrocotus solaris* 18cm

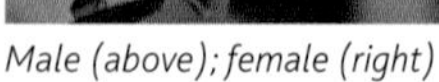

Male (above); female (right)

Males of this species and the tiny Small Minivet *Pericrocotus cinnamomeus* are orange rather than scarlet (unlike the Scarlet Minivet *Pericrocotus speciosus*). The Grey-chinned is distinguished by the grey (not blackish) cheeks and large orange-red wing patch. The female is yellow where the male is red, and unlike many other female minivets (including Scarlet) has no yellow on the forehead. Common resident of evergreen forests in the northern mountains above 1000m; the similar Small Minivet occurs below 1000m.

GREEN IORA *Aegithina viridissima* 14cm

The ioras are birds of the treetops, predominantly green or yellow. This species is largely dark olive-green above, with a conspicuous yellow eye-ring; wings and tail black, with a double white or yellowish wingbar. The breast of the male is green, grading into a yellow belly, with vent silvery-white. The female is entirely yellow below. Found in evergreen forest, and on the forest edge, up to 800m. Uncommon resident in the southern half of the peninsula. Unlike next species, not found in cultivated areas.

COMMON IORA *Aegithina tiphia* 15cm

The bright yellow underparts and double white wingbar distinguish this common species of gardens, open woodland and scrub throughout the country. Ioras and leafbirds keep to the canopy, or the upper branches of garden trees. Both sexes are olive-green above; the male often has black on the crown and mantle (breeding season). The loud double whistle, both syllables long, the second one noticeably lower, should soon become familiar. Very common resident.

GREATER GREEN LEAFBIRD *Chloropsis sonnerati* 21cm

Male (above); female (below)

Leafbirds are predominantly green, and spend much of their time in the upper canopy. The male of this species can be told from all other leafbirds by its combination of a black throat and lower face with a complete absence of yellow. The female is the only leafbird with a yellow throat, and the only one with a prominent yellow eye-ring. Both sexes have an inconspicuous blue moustache. Common resident in forest and mangroves throughout the peninsula ranging up to 900m.

GOLDEN-FRONTED LEAFBIRD *Chloropsis aurifrons* 20cm

The only leafbird with a bright golden-yellow forecrown (male only). The black throat patch is much larger than that of other species (except the distinctive Orange-bellied), and is deep blue in the centre. There is a pale blue patch on the shoulder area. The juvenile is green all over, with minimal yellow on the forehead and no blue (except a slight blue moustache). This species prefers mixed deciduous forests of the lower hills, and does not occur in the peninsula. Very common resident.

BLUE-WINGED LEAFBIRD *Chloropsis cochinchinensis* 19cm

This is the only species of leafbird in which both sexes have bluish primaries and tail and a blue patch at the carpal joint. Otherwise very similar to the Lesser Green Leafbird *Chloropsis cyanopogon*, though the yellow border to the male's black throat patch is broader and more diffuse. The female may show a bluish tinge to the green throat. A common resident of evergreen and mixed deciduous forests throughout the country, below 1200m.

ORANGE-BELLIED LEAFBIRD *Chloropsis hardwickii* 20cm

The male has a much larger black throat patch than other species, the black extending to the upper breast, and a brilliant orange-yellow belly and vent. The primaries are blackish, and there is a blue patch on the carpal joint. The female has the head and breast green, and the orange of the belly is duller than the male's. Like all leafbirds, this species imitates the songs of other species. It is found only in the north, and prefers evergreen forests above 600m. Fairly common resident.

BLACK-HEADED BULBUL *Pycnonotus atriceps* 18cm

This species' black head and predominantly yellow plumage, with yellow-tipped tail with black subterminal band, are easily recognized. Less conspicuous and aggressive than the following species. Bulbuls are medium-sized fruit-eating birds with a fairly strong bill; most species are not difficult to see, and many occur in small flocks. In the forest, look for them in fruiting trees. Common resident in forests and secondary growth throughout the country except in the higher hills.

BLACK-CRESTED BULBUL *Pycnonotus flaviventris* 19cm

Perhaps the most conspicuous resident bulbul in forests throughout the country. The black head and prominent crest, brilliant yellow underparts, and noisy and aggressive behaviour all demand attention. White eye stands out on dark head. The local subspecies in the eastern provinces has a red throat patch. There is only a trace of yellow on the wing, and none on the tail. Very common resident in mixed and evergreen forests at all altitudes.

SCALY-BREASTED BULBUL *Pycnonotus squamatus* 16cm

The bold scaly white marks on the black breast and flanks separate this from all other species of bulbul. Black head with contrasting white throat also distinctive. Wings bright yellow-olive with blackish primaries; rump and vent orange-yellow. Tail black, with outer feathers broadly tipped white. A bird of the treetops. Rather common resident of evergreen forests in the southern half of the peninsula, found on hill slopes up to 1000m.

RED-WHISKERED BULBUL *Pycnonotus jocosus* 20cm

The cheeky crested bulbul of urban areas, this species is also found wherever there are clearings in the forest. The vertical crest, thin black moustache and broad black smudge on the side of the neck are diagnostic; small red patch above white cheeks is often invisible. Upperparts dark brown; underparts white, with vent bright red. Occurs in many habitats, but not in dense forest. Extremely common and conspicuous resident except in the north-east.

SOOTY-HEADED BULBUL *Pycnonotus aurigaster* 20cm

A rather dark bird with pale grey uppertail-coverts, the latter conspicuous as it flies away from you. The crown and facial mask are black. Underparts greyish-white, with the vent either red or, in some populations (especially in the west), lemon-yellow. This species is very common in the north, the central plains and the north-east, but does not extend to the peninsula. It may be found in the same areas as the Red-whiskered Bulbul, but is not usually seen as close to human habitations.

STRIPE-THROATED BULBUL *Pycnonotus finlaysoni* 19cm

This bulbul's yellow streaks on the forehead and throat are unmistakable. Dull greenish-brown above and on crown, with brighter edges to primaries and tail feathers. Underparts dull greyish, with bright yellow undertail-coverts. This species is a common resident in forested areas throughout most of the country; it tends to prefer clearings or the forest edge, also occurring in scrub and secondary growth, up to about 900m.

FLAVESCENT BULBUL *Pycnonotus flavescens* 22cm

This common montane species is best distinguished by the short white supercilium in front of the eye, bordered below by a blackish loral stripe. Head and throat are otherwise greyish, with an insignificant crest. Chiefly olive-brown above and yellowish below, brightest on the undertail-coverts. Juveniles have a yellower head with less obvious supercilium. One of the commonest bulbuls resident in the northern forests, where it inhabits more open areas above 900m.

YELLOW-VENTED BULBUL *Pycnonotus goiavier* 20cm

The only bulbul with a very broad and complete white supercilium, black lores and yellow vent. Also has white throat and blackish central crown; upperparts and tail dark brown. Young birds are less well marked. This is one of the commonest bulbuls of cultivated areas, including palm groves, south of the latitude of Bangkok, preferring coastal areas or localities with lakes and ponds; also occurs in secondary growth and scrub. Resident.

RED-EYED BULBUL *Pycnonotus brunneus* 19cm

At close range, the red eye without orbital ring is diagnostic. This is another dull brown forest bird, showing very few relieving features in its uniform plumage; underparts only a little paler and greyer than upperparts. The best guide to its identity is probably the voice, a high-pitched bubbling or trilling sound, rising distinctly towards the end. Common resident in and on edges of forests and secondary growth of the extreme south.

MOUNTAIN BULBUL *Ixos mcclellandii* 24cm

The white streaks on the throat and crown separate this species from other bulbuls of the higher mountains. The crown is dark brown and has a slightly shaggy appearance, and the throat is paler and greyer; sides of head and upper breast rusty-brown. Upperparts, wings and tail olive, tinged yellow. Belly whitish, with yellow vent. Common resident of evergreen forests in the north, from 800m right up to to the mountain summits.

ASHY BULBUL *Hemixos flavala* 20cm

The only bulbul, apart from the distinctive Black-headed, with an obvious yellow patch on the closed wing (though this is lacking in the race inhabiting the peninsula). Head dark brown, with a distinct crest; throat, belly and vent white, rest of underparts grey. Dull chestnut patch behind the eye. The distinctive call sounds like a bicycle bell. A common montane species resident in evergreen forests in the north and south-east, and at lower elevations in the extreme south.

(HIMALAYAN) BLACK BULBUL *Hypsipetes leucocephalus* 25cm

An all-black bulbul with bright red bill and legs and a shaggy head. Two wintering subspecies have a white head in adult plumage, one also having a white upper breast. Often occurs in much larger flocks than most bulbuls. Flocks keep to the canopy, and may often be located from a distance by their constant chattering and squawking calls. Restricted largely to the montane forests of the north and west, where it is a common resident, though winter visitors also reach the south-east.

BLACK DRONGO *Dicrurus macrocercus* 28cm

A glossy black bird with a deeply forked long tail. Normally found in open country (other drongos are forest birds). Frequently associates with cattle, feeding on insects disturbed by them; perches on telegraph wires, and often flies out to catch insects on the wing. Extremely common in all kinds of open habitats, including in villages and around towns, throughout the country, though it reaches the peninsula only in winter. Mainly in lowlands, but passage migrants also seen at higher elevations.

ASHY DRONGO *Dicrurus leucophaeus* 29cm

A dark grey bird of forest clearings. It has a deeply forked tail like the Black Drongo, but is never as dark. One wintering race is pale grey and has a large white cheek patch. More slender than the cuckoo-shrikes with similar colouring; it perches characteristically on exposed branches. Common resident in the north and some other areas (in the peninsula, occurs only in mangroves and coastal scrub). It is a common winter visitor throughout the country (when it often moves into cultivated areas).

LESSER RACKET-TAILED DRONGO *Dicrurus remifer*

25cm + 35–40cm tail

The racket-tailed drongos both have elongated tail feathers with a 'racket' on the end. They are birds of the middle storey of the forest, the Lesser being more or less confined to the north and west. If the rackets are present (they are often broken off), the two species may easily be distinguished by the shape of the racket: on Lesser, in contrast to the Greater, it extends both sides of the feather spine. The Lesser also has a small tuft of feathers covering the base of the upper mandible, giving it a markedly 'flat-headed' silhouette. It is commonly found in bird waves. Resident.

GREATER RACKET-TAILED DRONGO *Dicrurus paradiseus*

32cm + 30cm tail

This species is best distinguished from Lesser Racket-tailed Drongo by the shape of the tail racket, which is restricted to one side of the feather spine only and is slightly twisted. It also has a heavier bill and much more prominent crest. It has a variety of loud whistling calls, and imitates several other species. It is a common experience to track down the source of an unfamiliar call, only to find this species making it! Resident through most of the country, but absent in much of the north-east.

HAIR-CRESTED DRONGO *Dicrurus hottentottus* 32cm

A drongo with an extraordinary lyre-shaped tail, the tips of which are slightly, but very noticeably, upturned. Plumage black, with a glossy metallic green sheen. Long, wispy crest often not visible in the field. Has long, slightly decurved bill. More commonly found in flocks than other drongos. Call like a creaking rusty gate-hinge. A common resident of forests and secondary growth on the lower mountain slopes, but in some areas a winter visitor only; absent altogether from the peninsula.

BLACK-NAPED ORIOLE *Oriolus chinensis* 27cm

Most orioles are brilliant yellow and black birds with characteristic musical calls. The adult male of this species is yellow, with a broad black band through the eye and across the nape, and some black on the wings and tail; the female is duller, and streaked below. As well as its fluty calls, it also gives a harsh, rather protracted, nasal call. Common winter visitor throughout the country, usually in the forest canopy, but also in open country with scattered trees.

BLACK-HOODED ORIOLE *Oriolus xanthornus* 25cm

A bright yellow bird with a black head and throat, and black on the wings and tail; immatures similar but duller. The only species with which it could possibly be confused is the Dark-throated Oriole *Oriolus xanthonotus*, which also has a black head and throat, but has black streaking on white underparts, and is restricted to the lowland forests of the peninsula. A common resident of forest areas up to about 800m, and in mangroves in the south.

ASIAN FAIRY BLUEBIRD *Irena puella* 25cm

The male has brilliant deep shining blue upperparts, including crown, rump and uppertail-coverts, with the rest of the plumage black apart from deep blue vent. The female is dull blue all over. Much bigger than even the largest of the blue flycatchers. It may be found in flocks or small parties, and quite often perches on exposed branches; attracted to fruiting trees. Common resident of evergreen forest, less numerous in mixed deciduous forest, in north and west.

EURASIAN JAY *Garrulus glandarius* 33cm

This is a large noisy bird of the canopy, with a distinctive white rump in flight. At rest, the white head, with black crown and moustachial stripe, contrasts with the milk-chocolate back and black wings and tail. The blue patch on the wing is more obvious in flight. The hoarse raucous call is distinctive, and it utters a variety of other calls (some mimetic). Occurs in small flocks. A common resident of more open forests of the north and west.

COMMON GREEN MAGPIE *Cissa chinensis* 38cm

This splendid apple-green bird mixes with laughingthrushes in the lower storey of the forest. The wings are chestnut-red, with tertials tipped black and white, the red colour being very evident in flight. There is a black eye-stripe, and the bill is bright red. Often travels in small parties, which herald their arrival with loud screeching calls. Nevertheless, this is a shy species, often difficult to see. A common resident of the forests mainly of the north and west and also in Khao Yai.

SOUTHERN JUNGLE CROW *Corvus macrorhynchos* 51cm

Noticeably heavy bill, large size and all-black coloration. Difficult to distinguish from Large-billed Crow (*C. japonensis*)and Eastern Jungle Crow (*C. levaillantii*). Broad wings with spread primaries and slow wingbeats are diagnostic in flight, which is direct and purposeful. Habitually calls in flight, a low-pitched, harsh 'kaak'. Occurs in a broad range of habitats, from woodland and mangroves to open country, and may even be seen in cities and towns.

YELLOW-CHEEKED TIT *Parus spilonotus* 14cm

This bird's erect black crest together with its acrobatic antics so typical of tits make it easy to recognize. The cheeks and forecrown are yellow, brighter on the male, which also has a black throat, central breast and belly and yellow flanks. The female is much duller below. The wings of both sexes are black, heavily marked with white. Utters a ringing 'tee-cher tee-cher' call. A common resident of the montane forests of the north, above 900m, a habitat it shares with the better-known Great Tit.

SULTAN TIT *Melanochlora sultanea* 20cm

A large spectacular tit with a bright yellow crest. The face and the entire upperparts, including wings and tail, together with the throat and upper breast are black, with the remaining underparts bright yellow. The black areas are duller on the female. This species is usually found in the canopy of evergreen forest, but also in the upper branches of smaller deciduous trees. Fairly common resident in the north and west of the country, and in the peninsula, up to 1000m.

VELVET-FRONTED NUTHATCH *Sitta frontalis* 12cm

Nuthatches are the only birds which habitually walk both up and down the trunks of trees. The most distinctive features of this species are the white throat and bright red bill. It has a black patch on the forehead and is violet-blue above, compared with the more grey-blue tone of other nuthatch species; unlike the latter, its ventral area is plain, without pale scallop markings. Follows bird waves. Very common resident in the forests of the north and west, and in the peninsula.

SHORT-TAILED BABBLER *Malacocincla malaccense* 15cm

The very short tail is normally enough to identify this babbler at first glance. A similar species with this feature is the thicker-billed Horsfield's Babbler *Trichastoma sepiarium*, but that is rare, and occurs only in one small area in the extreme south. The Short-tailed also has a black moustachial stripe and pinkish legs. Found in evergreen and secondary forests up to 900m. Fairly common resident in the southern half of the peninsula.

SOOTY-CAPPED BABBLER *Malacopteron affine* 17cm

The combination of uniformly dark crown and lack of a moustachial stripe distinguishes this babbler from a number of confusingly similar species. The upperparts are dull brown, becoming rufous on the rump and tail; the underparts are whitish. The similar, and much commoner, Moustached Babbler *Malacopteron magnirostre* has, as its name suggests, a moustachial stripe, and the crown is not noticeably darker than the back. Both are resident in the lowland forests of the peninsula.

RUFOUS-CROWNED BABBLER *Malacopteron magnum* 18cm

A relatively distinctive babbler with a chestnut forecrown and a black patch on the nape. The underparts are white, with grey streaks on the breast. It is a scarce resident of the evergreen forests of the extreme lowlands in the peninsula. The very similar Scaly-crowned Babbler *Malacopteron cinereum* occurs in the same areas, but ranges higher up the hills; it is a smaller-looking bird, with pinkish (not grey) legs, and lacks the grey streaking on the breast.

STREAKED WREN-BABBLER *Napothera brevicaudata* 13–17cm

A largely ground-dwelling species with conspicuous pale streaks or scale-like markings on the crown, throat and upper back. Underparts below throat reddish-brown. This species prefers moist gullies and rocky outcrops, and even limestone hills outside the range of the rather similar but larger Limestone Wren-babbler *Gypsophila crispifrons*. Fairly common resident of evergreen forests in parts of the north, the south-east and the peninsula.

GOLDEN BABBLER *Stachyridopsis chrysaea* 13cm

The bright golden underparts and black-streaked golden crown identify this tiny babbler. The rest of the upperparts are golden-olive, greener on the wings and tail. Black lores. Easy to confuse with the small warblers, but does not have either prominent eye-ring or supercilium. Generally in the lower branches of the trees or bushy undergrowth, not on the ground like many babblers. Common resident in the mountain forests of the north, above 900m, and on the mountains of the peninsula.

GREY-THROATED BABBLER *Stachyris nigriceps* 15cm

A regular member of mixed-species feeding parties (bird waves) of evergreen forests, where it keeps to the undergrowth. It can be identified mainly by the head pattern: the crown streaked with white, a bold black stripe above the eye, a whitish stripe behind the eye, a broad white moustache, and a grey throat. The rest of the plumage is olive-brown, paler below. A characteristic and common resident of northern montane forests; also found in the peninsula.

PIN-STRIPED TIT-BABBLER *Macronous gularis* 13cm

Another regular member of bird waves, this species prefers the lower branches of the trees. The combination of chestnut crown, yellow supercilium, and black streaks on the yellow throat and upper breast is diagnostic. Upperparts olive with brown tinge. Found in deciduous and evergreen forest, as well as in secondary growth and scrub, occurring to 1500m. Common to very common resident throughout the country, except in the central plains and most of the north-east.

WHITE-CRESTED LAUGHINGTHRUSH *Garrulax leucolophus* 30cm

A striking bird, with snow-white crest, throat and breast contrasting with the black mask and bright rufous upperparts and lower belly. The commonest laughingthrush of lowland forest (up to 1200m) in the north and west and in some parts of the north-east. Its cackling calls and hysterical laughter are frequently heard, but this species is not too easy to see except in some places where it is accustomed to man (e.g. national park headquarters). Common forest resident.

LESSER NECKLACED LAUGHINGTHRUSH *Garrulax monileger*

30cm

The two necklaced laughingthrushes are immediately separated from all others by the striking black 'necklace'. Both have a white supercilium, and a long black eye-stripe joining up with the necklace. This species has unmarked pale ear-coverts and throat, whereas Greater Necklaced Laughingthrush *Garrulax pectoralis* has a black moustache bordering black-streaked white ear-coverts. Both species are common forest residents up to 1200m, the Greater only near the western border; both absent from the peninsula.

WHITE-NECKED LAUGHINGTHRUSH *Garrulax strepitans*

30cm

A dark laughingthrush of the deep evergreen forests. The crown is brown, and the face, throat and upper breast black. There is a small chestnut spot on the ear-coverts, behind which is a white patch which merges into an ill-defined greyish collar. Otherwise, dark grey-brown above, greyish below, and generally appears dark. Occurs in small flocks. Rather uncommon resident in the northern montane forests up to 1800m.

BLACK-THROATED LAUGHINGTHRUSH *Dryonastes chinensis*
30cm

A relatively distinctive species, having bright white cheeks contrasting with a black bib and dark grey crown. Plumage is otherwise mainly dark greyish-olive. Has perhaps the best song of all the laughingthrushes, which are generally noisy rather than tuneful birds: gives repeated mellow whistles and high-pitched squeaks. Found in mixed flocks with the White-crested and Lesser Necklaced Laughingthrushes in the forests of the north and west, and in Khao Yai. Fairly common resident.

CHESTNUT-CROWNED LAUGHINGTHRUSH
Garrulax erythrocephalus 27cm

The bright yellow-olive wings and tail and chestnut crown immediately separate this from all other laughingthrushes. The facial mask is black and the cheeks silvery-grey. The wing shows a bright chestnut patch (tips of greater coverts) with a black patch (primary coverts) directly behind it. The back and underparts are olive-grey. It prefers thick undergrowth. A fairly common resident of high montane evergreen forests in the north and in the peninsula.

BROWN-CHEEKED FULVETTA *Alcippe poioicephala* 16cm

The fulvettas are small brown undergrowth-haunting birds commonly found in bird waves. This species is identified by the combination of the pale grey crown and a long black stripe from the bill passing over the eye, and reaching the upper back. The front part of this black line is reduced or absent in more southern races. The cheeks are pale brown. Otherwise brown above, buffy below. A common resident in forests, bamboo, and secondary growth up to 1100m, in the north and west of the country.

WHITE-BELLIED ERPORNIS *Erpornis zantholeuca* 13cm

Resembling a leaf-warbler at first glance, this species can be identified by its prominent crest and lack of a supercilium. It is yellowish-green above and white below, with yellow vent. At close range, the decurved pink bill is visible. It frequently travels in bird waves, when it occupies the middle layer of the forest, at 3–6m above the ground. Common resident in forested areas throughout the country, except on the highest mountains; absent from the central plains and most of the north-east.

WHITE-BROWED SHRIKE-BABBLER *Pteruthius flaviscapis* 17cm

A montane species which can easily be recognized by its habit of walking along horizontal branches near the canopy. It usually draws attention by its frequently repeated and very monotonous four-note call, 'chi-chewp, chi-chewp'. The male is largely black above, with a white supercilium, and a golden patch on the tertials; the female is greyer, with olive wings and tail. Very common forest resident on all the higher mountains (above 800m).

BAR-THROATED MINLA *Chrysominla strigula* 17cm

This brightly coloured species is noisy and gregarious, and unafraid of man. The predominant impression is of a bright yellow-buff bird with black and white bars on the throat. The wings show a large chestnut-orange patch, the tertials are black with broad whitish edges, and the centre of the tail is chestnut; the crown is golden-chestnut and the back olive-green. Found only on the upper slopes of Doi Inthanon, where it is a characteristic and common forest resident from 2000m to the summit.

SILVER-EARED MESIA *Mesia argentauris* 18cm

This attractive bird's black cap and moustache, combined with white ear-coverts, may remind the observer of an outsized Great Tit *Parus major.* The rest of the plumage, however, is greyish-green above, and bright golden-orange below and on the nape, with red patches on the wing (both sexes) and the uppertail- and undertail-coverts (male only). Prefers thick undergrowth or scrub. Common resident in the higher mountains of the north, above 1300m.

SPECTACLED BARWING *Actinodura ramsayi* 24cm

The bold white eye-ring and the conspicuous narrow black barring on chestnut to brown wings and long tail combine to produce an impression unlike that of any other species. The general colour of the crown and underparts is a rich yellowish-buff, with the upperparts a darker olive; lores black, increasing prominence of white eye-ring. Common resident of forest, scrub and secondary growth on a few higher mountains in the north; it may be seen by the roadside on Doi Inthanon, up to about 2100m.

DARK-BACKED SIBIA *Malacias melanoleusus* 23cm

A bird of the canopy, black above and white below, with a long tail (though not so long as that of the Long-tailed). In flight, it reveals a small patch of white on the wing and white tips to the tail feathers. Usually seen in small flocks. More likely than the following species to come down from the canopy to feed on berries in low vegetation. Common resident in the evergreen forests of the northern mountains, from 1000m up to the summits.

LONG-TAILED SIBIA *Heterophasia picaoides* 30cm

A large, dull grey bird with a tail longer than its body. Similar in shape to the previous species, but grey above and below, with a large white wing patch. Tail feathers tipped with white. This species is found in the same areas as the Dark-backed Sibia, but tends to keep to the canopy, where small flocks can frequently be seen feeding on flower nectar. Common resident of forest and secondary growth in the northern mountains area occurring between 900m and 1800m.

GOLDEN-BELLIED GERYGONE *Gerygone sulphurea* 9cm

A tiny and rather nondescript warbler. Whitish lores, but no superciliary stripe; grey-brown upperparts, and yellowish breast and underparts. White subterminal spots on tail feathers. This species is commonly found in mangroves and coastal scrub, but occurs also in other lowland habitats (forest, plantations, secondary growth). Common resident in the southern peninsula, and on the coast further north, though absent from the south-east region.

ARCTIC WARBLER *Phylloscopus borealis* 13cm

This is usually the first leaf-warbler to pass through in autumn, and the last in spring. There are several very similar species in this difficult genus. Arctic Warbler usually has a single wingbar (not shown in photograph), and is greener above and whiter below than the Dusky Warbler *Phylloscopus fuscatus*. It tends to keep high in the trees. Inhabits mixed and evergreen forests, as well as mangroves, secondary growth and sometimes gardens. A very common passage migrant in the north, and a winter visitor in the south.

YELLOW-BROWNED WARBLER *Phylloscopus inornatus* 11cm

This tiny warbler can be found almost everywhere in winter. It has a pronounced pale supercilium and a double wingbar, but no central crown-stripe. It is usually very active, often travelling in small parties or with other species, but may also be seen on its own. There are several very similar species, but they all have a central crown-stripe, and some also have a yellow rump. Abundant winter visitor, occurring anywhere with trees.

ORIENTAL REED WARBLER *Acrocephalus orientalis* 20cm

This is the large brown warbler of the reedbeds. Brown above and buffy below, with a prominent pale supercilium and a relatively heavy bill. The supercilium helps to distinguish it from the even stouter- but shorter-billed Thick-billed Warbler *Acrocephalus aedon*, which shares the same habitats but also occurs in drier areas. Although it usually lives in reedbeds, it may also be found in scrubland, but usually close to water. Common winter visitor throughout the country.

BLACK-BROWED REED WARBLER *Acrocephalus bistrigiceps* 14cm

The black eyebrow over the whitish supercilium distinguishes this warbler from similar species. It is normally very skulking, but frequently utters its very harsh call (much hoarser than that of the Oriental Reed Warbler). The best time to see this species is just after rain, when it often comes to the top of the reeds to dry itself. Inhabits reedbeds and tall grass and scrub in wet or marshy places. Common winter visitor in most of the country but absent from most of the north-east (except near the Mekong).

ZITTING CISTICOLA *Cisticola juncidis* 11cm

A tiny, heavily streaked warbler which likes to perch on the tops of the grass stems. Its flight is weak and undulating, and in territorial display it repeatedly gives a ticking call while flying. The very similar Bright-headed Cisticola *Cisticola exilis* has a golden crown (breeding male) and different calls; it is also less widespread, and prefers taller grass. The Zitting Cisticola can be found anywhere in open grassland throughout most of the country. Very common resident.

RUFESCENT PRINIA *Prinia rufescens* 12cm

Prinias are long-tailed grass-frequenting warblers, often seen holding the tail cocked. This species has a white supercilium extending beyond the eye, and rufous primaries. The head is grey, becoming browner outside the breeding season; the underparts are buffy, with a whitish throat. It prefers undergrowth in open forest or grasslands, usually not much above 500m, but sometimes ascending to 1600m. Very common resident throughout the country, except in the central plains.

YELLOW-BELLIED PRINIA *Prinia flaviventris* 13cm

The yellow belly and vent of this species in the breeding season are diagnostic. The head is grey, with a whitish throat. It has a narrow, indistinct white supercilium extending only as far back as the eye. The song, transcribed as a short and musical 'didli-idli-u' (David Wells), is heard incessantly in the grassy lands and scrub which it prefers; in the south, it may also be found on the landward side of mangroves. Very common resident, mainly in the lowlands, but also up to 800m throughout the country.

PLAIN PRINIA *Prinia inornata* 15cm

This is a larger and browner bird than the Yellow-bellied Prinia, with buffy underparts. The supercilium, white to buff in colour, is better marked than on most prinias and extends beyond the eye. No rufous on wings. This species seems to prefer wetter districts than the Yellow-bellied, being found in vegetation of paddyfields, marshes, ponds and similar places, including edges of mangroves. A very common resident throughout much of the country except the peninsula and southern parts of the west.

COMMON TAILORBIRD *Orthotomus sutorius* 12cm

Both the loud repetitive calls and the sharply cocked tail are characteristic of all tailorbirds. This species is the common tailorbird of gardens and open country throughout Thailand. It looks like a small warbler with a chestnut crown and long, graduated tail. The upperparts are greenish and the underparts whitish, often with some dirty grey spots on the throat. It is resident anywhere where the country is fairly open and usually fairly dry.

DARK-NECKED TAILORBIRD *Orthotomus atrogularis* 11cm

The black patch on the side of the neck of the male is distinctive, though lacking on the female. The plumage is otherwise similar to that of the Common Tailorbird but much yellower-looking, with brighter yellow-green upperparts and yellow undertail-coverts and flanks; the chestnut on the crown also extends farther back, reaching well onto the nape. This species prefers thicker undergrowth, forest and wetter country than the Common Tailorbird. Very common resident throughout the country up to 1200m.

ASHY TAILORBIRD *Orthotomus ruficeps* 12cm

A dark grey tailorbird with rufous crown and sides of head. In all plumages, the rufous extends noticeably below the eye, covering the ear-coverts and the cheeks; on adult males it also extends over the chin. Black and white tips to the tail feathers. Females and young birds have varying amounts of white on the underparts. Prefers mangroves or coastal scrub, also swampy forest. Fairly common resident, but restricted to the southern half of the peninsula.

MOUNTAIN TAILORBIRD *Phyllergates cuculatus* 12cm

The dark stripe through the eye and the white supercilium separate the adult from other tailorbirds and from the rather similar Chestnut-crowned Warbler *Seicercus castaniceps*. It has a whitish throat and breast, both bordered with grey at the sides, and bright yellow lower underparts. Bill noticeably long. Juveniles are even more warbler-like, with dull olive upperparts, pale supercilium, and pale yellow underparts. Does not usually cock its tail. Fairly common local resident in the montane forests of north, especially where there is bamboo.

WHITE-BROWED SHORTWING *Brachypteryx montana* 15cm

Male (above); female (right)

A small blackish-looking bird with conspicuous white supercilia, these sometimes appearing to join across forehead. This dark coloration, which only the male possesses, is in fact deep blue; the female is dark brown, with rufous forehead and short supercilium. A characteristic species of the upper slopes of Doi Inthanon and other high mountains in the north. It keeps to the forest floor, and may often be seen on the path to the boardwalk at the top of Doi Inthanon; not at all shy. Resident.

BLUETHROAT *Luscinia svecica* 15cm

A typical view of this species is of the chestnut outer tail feathers as it flies away. The male has a mainly blue throat with a red central patch in breeding plumage; the blue is bordered by a black gorget and then a further band of red. The female has a blackish breast band of spots on buffy underparts. Skulking. It frequents marshy undergrowth in the plains, often wetter areas than used by Siberian Rubythroat *Luscinia calliope*, and does not extend up the mountains. Common in winter in the northern rice paddies, the north-east and central plains.

ORIENTAL MAGPIE-ROBIN *Copsychus saularis* 23cm

This striking black and white bird is the common thrush of gardens and reasonably open places, and will be seen throughout the country. It is conspicuous and not afraid of man. The female is grey on the head and breast where the male is black. It frequently perches with its tail cocked. Often uses trees and telegraph wires as a perch, and gives a loud but pleasing whistled song. Very common resident in open woodland, town gardens and cultivated land everywhere across the region, up to 1800m.

WHITE-RUMPED SHAMA *Copsychus malabaricus* 22cm

This is one of the finest songsters of all Thai birds, but is as skulking as the Magpie-robin is conspicuous. It is black, with deep orange underparts, and a large white patch on the rump; the outer tail feathers are also white. The male has a greatly elongated tail. Keeps to thick cover, from where the rich, fluty, melodious song may be heard. A very common resident in forest situations throughout the country; absent from the Bangkok area and much of the north-east.

CHESTNUT-NAPED FORKTAIL *Enicurus ruficapillus* 20cm

Forktails are black and white birds, somewhat resembling large wagtails with long and deeply forked tails. They live on the edges of forest streams. This species has black scaly markings on the breast, and a rich chestnut crown, nape and upper mantle (on females, the chestnut extends over the entire mantle and back). No other Thai forktail has chestnut in the plumage. Found by stony streams in evergreen forests from the extreme west southwards through the peninsula. Uncommon resident.

EASTERN STONECHAT *Saxicola maurus* 14cm

An upright-perching bird of open country. The male is distinguished by its black head, white sides to the neck, and orange underparts. The upperparts are blackish, with a whitish rump and white wing patch. Females and immatures are browner, with buffy rump, and without black on the head. Plumages of wintering birds vary considerably. Like all chats, it perches on the tops of small bushes and in other exposed places, but never far above the ground. Very common in winter; resident on a few northern mountains.

GREY BUSHCHAT *Saxicola ferreus* 16cm

The grey chat of open forest. The male is grey, with a black face and white supercilium and throat. The female is brown. The distinct supercilium distinguishes both sexes from all other chats (female Stonechat has a trace of a supercilium, but nothing like as distinct). The female is also unstreaked (Stonechat is heavily streaked above in all plumages). Occurs in clearings and woodland edges; much more commonly found in forest areas than the other chats. Common winter visitor, mainly in the north; also breeds on a few high mountains.

BLUE WHISTLING-THRUSH *Myophonus caeruleus* 33cm

A large, blackish-looking thrush, often seen flying along mountain streams or perched on a rock beside them. It is also quite often seen at the roadside in forest areas. Its plumage is deep violet-blue with lighter spots and streaks. The bill is usually yellow (but can also be black). As it alights, it fans its tail, and this action is also seen when the bird is alarmed. Generally shy, and flies away if approached. Common resident in forested areas; absent in the central plains and most of the north-east.

ASIAN BROWN FLYCATCHER *Muscicapa dauurica* 13cm

The upright stance and alert air of this rather dull-coloured species soon become familiar. Brown to greyish above, with a slight white wingbar (only in fresh autumn plumage), and off-white below. Whitish eye-ring. Whitish area in front of the eye (wintering race). Usually perches on a branch just below the canopy. The common brown flycatcher of the more open woodlands, found throughout Thailand in winter, and resident on the northern mountains.

LITTLE PIED FLYCATCHER *Ficedula westermanni* 12cm

This striking black and white bird cannot be confused with any other flycatcher. The male is black above, white below, with a broad white supercilium and wing patch and a white basal patch on each side of the tail. The female is rather nondescript brown above, whitish below. The most likely confusion is with the Bar-winged Flycatcher-shrike, which does not have the white supercilium. Keeps to the higher branches of the trees. A common resident of the mountain forests of the north.

GREY-HEADED CANARY-FLYCATCHER *Culicicapa ceylonensis* 13cm

This beautiful yellow bird with a grey head is the most conspicuous of the resident forest flycatchers. The back is dark yellow-olive, the head and upper breast grey, and the underparts bright yellow. Very active and noisy, 'churring' as it makes frequent feeding sallies in the middle storey of the forest, and often accompanying bird waves. Common resident of the north, west and peninsula; common winter visitor in most of the rest of the country, occurring also in well-wooded gardens.

BLUE-AND-WHITE FLYCATCHER *Cyanoptila cyanomelana*
18cm

A blue flycatcher with a blue-black breast and sharply contrasting white underparts, and with white basal patches on each side of the tail. The female is greyish-brown above, browner on wings and tail, with a rufous-buff breast and white underparts. It usually prefers the middle storey of the forest, 3–5m above the ground. Found in evergreen forests but may be seen in any wooded areas on passage. It is an uncommon passage migrant in most parts of the country; absent from the north-east.

LARGE NILTAVA *Niltava grandis* 21cm

A rather large, dark blue flycatcher, appearing blackish in poor light. It has bright iridescent blue patches on the crown, shoulders and rump and just below the eye. The female is rusty-brown, tinged olive on the upperparts, with a blue spot below the eye and a buff throat. Immatures also have buff-speckled upperparts. Prefers the middle storey of the forest; most niltavas prefer the undergrowth. A mountain bird, resident in the evergreen forests above 900m, mainly in the north of the country.

Male (above); female (below)

VERDITER FLYCATCHER *Eumyias thalassinus* 17cm

A distinctive bird that is easily recognized by its unusual plumage colour. Powdery bluish-green above and below, with a black line joining bill and eye. The female is similar but duller. Some white on the under tail-coverts. The apparent colour of the plumage, especially that of the male, changes according to the light. It prefers the canopy, often sitting on an exposed perch from which it sallies to catch insects. Common resident and winter visitor in evergreen forest areas, except in the north-east.

BLUE-THROATED FLYCATCHER *Cyornis rubeculoides* 15cm

The dark blue throat distinguishes the male from two similar species which are also common in the same areas: Tickell's Blue Flycatcher *Cyornis tickelliae* (orange throat sharply divided from white underparts), and Hill Blue Flycatcher. Note that Blue-throated sometimes shows a narrow wedge of orange extending up towards the bill. Female has blue areas replaced by brown, with rufous-tinged tail. Resident in deciduous forest and bamboo in the north and west; winters in peninsula, in lowland evergreen forest.

HILL BLUE FLYCATCHER *Cyornis banyumas* 15cm

This is the commonest of the blue flycatchers in the northern forests. The plumage is blue above, with a blackish patch through the eye. The breast is orange, grading into white underparts. The female is very like female Blue-throated, but has a richer orange throat and breast more sharply demarcated from its cheeks. Hill Blue is a common resident of the middle storey of the forest in most forested areas, though less widely distributed in the peninsula; absent from most of the north-east and central plains.

WHITE-THROATED FANTAIL *Rhipidura albicollis* 19cm

Fantails are dark grey-brown flycatchers of the forest undergrowth, characterized by their habit of cocking and fanning the tail. This species is all dark, except for white throat, supercilium and tips to the outer tail feathers. Common resident of the evergreen mountain forests of the north and west, above 600m. The White-browed Fantail *Rhipidura aureola*, which occurs in most of the same areas, is largely white below, and is normally found below 900m.

PIED FANTAIL *Rhipidura javanica* 18cm

This is the common fantail of the lowlands. It is dark above and white below, with a broad black breast band. The head has a thin white supercilium and the outer tail feathers are broadly tipped white. Calls with harsh churrs and whistling song. It is found in woodland, gardens, scrub and mangroves, showing a preference for the proximity of water. It is a common resident throughout the country except in the north and west; the ranges of this and White-throated scarcely overlap.

BLACK-NAPED MONARCH *Hypothymis azurea* 17cm

A predominantly blue flycatcher which looks almost black in the dim light of the middle storey, or up near the canopy. At close range, the male is dull blue, with a small black crest and a narrow black gorget; the belly is white. The female is browner and lacks the black markings; the blue is duller, and is restricted to the head. Inhabits forests and open woodland, extending into mangroves and gardens in winter. Very common resident throughout the country up to about 1200m.

ASIAN PARADISE-FLYCATCHER *Terpsiphone paradisi* 21cm

Male (white morph)

Female

The elongated central tail feathers of the male can add a further 23cm to its length. Females (which lack extended tail feathers) and most males are bright chestnut above, with the head dark grey, almost black, this colour becoming lighter on the breast and merging into the white belly. Some males are all white, apart from a black head and throat. Found in forests and secondary growth, in winter also in gardens and mangroves. Common resident and winter visitor throughout the country except the north-east.

BROWN SHRIKE *Lanius cristatus* 20cm

A compact bird with a thick hooked bill and a black eye-stripe. Characteristically perches in exposed places on low bushes or fences. Plumages are variable according to the time of the year and the subspecies involved, but most have a brown crown and a white supercilium (a clear white supercilium is diagnostic). The adult of one race has a pale grey crown, but can be distinguished by the pale rufous-buff underparts. Occurs in open areas. A common winter visitor throughout the country.

TIGER SHRIKE *Lanius tigrinus* 19cm

Very similar to the Brown Shrike, but has no white in the wings or tail. Bigger-billed and shorter-tailed than Brown Shrike. The adult male is distinctive: upperparts chestnut, narrowly vermiculated with black; head and nape grey, with black mask but no white supercilium; underparts white. A lowland species, preferring the forest edge, deciduous forest, or cultivated land. Fairly common passage migrant in most of the country except the north-east and south-east.

LONG-TAILED SHRIKE *Lanius schach* 25cm

The long blackish tail gives this species a different silhouette from that of other shrikes when it perches, as it often does, on telegraph wires. The adult's head is black, with the back pinkish-chestnut; throat and underparts white, washed with rufous on the flanks and becoming deeper rufous buff on vent. In flight, it shows a white wing patch. Juveniles have a barred ashy-grey crown, and are most easily distinguished by the silhouette. Common resident in open country in the north and the central plains.

ASHY WOOD-SWALLOW *Artamus fuscus* 18cm

Wood-swallows are like heavily built swallows with a broad base to the wing. The plumage is dark grey, paler below, with whitish uppertail-coverts. They are most often seen feeding in flight; the wings are much broader than those of true swallows, and the tail is square-ended. Inhabit open areas with some trees. They perch, often in noisy groups, on exposed branches of trees and telegraph wires. Common resident throughout the country (mainly lowlands) except the peninsula.

ASIAN GLOSSY STARLING *Aplonis panayensis* 20cm

The glossy black plumage and distinctive starling flight, with pointed wings and direct route, immediately distinguish this species from all others. The black plumage is glossed greenish. Young birds have white underparts heavily streaked with black, and the upperparts are brown-tinged and far less glossy. Red eye at all ages. Common resident in the peninsula, where it prefers cultivated areas and gardens, also roadsides, including in towns and villages.

ASIAN PIED STARLING *Gracupica contra* 24cm

A black and white starling, similar to but smaller than the Black-collared Starling. The easiest distinction is the entirely black chin and throat. Note also the white cheeks and forecrown, and the yellow bill with orange base. In flight, the underwing-coverts are white (black on Black-collared Starling). Found in open country and cultivated areas, including around towns, but usually avoids dry localities. Common lowland resident in the north and central areas.

BLACK-COLLARED STARLING *Gracupica nigricollis* 28cm

A large pied starling with a whitish head and black collar. Similar to but larger than the Asian Pied Starling, from which adults are best separated by the white throat contrasting with the broad black collar. Juveniles have the head and breast washed brownish-grey and lack the black collar. In the northern part of its range, this species is far more numerous than the Asian Pied. Very common resident throughout the country except in the extreme south, preferring urban, cultivated or deforested areas.

COMMON MYNA *Acridotheres tristis* 25cm

The familiar and common brown myna of the towns and villages. The head and upper breast are very dark brown, and the body dull chocolate-brown with white lower belly and vent. White wing patch. Bill and facial skin yellow. In flight, shows white underwing-coverts. Often associates with White-vented Mynas. May be seen commonly anywhere in open country, as well as in and around human habitations, up to 1500m. Very common resident throughout the country.

WHITE-VENTED MYNA *Acridotheres grandis* 25cm

This must be the commonest of all mynas, and one of the most conspicuous birds anywhere within its range. It is mainly black, with a prominent wispy crest on the forecrown. The vent, the tip of the tail, and a large wing patch are white, the latter conspicuous in flight. Bill and legs yellow. An open-country inhabitant, mainly in the lowlands, where often seen with cattle; also frequents gardens, including in urban areas. Very common resident, but absent from most of the peninsula.

HILL MYNA *Gracula religiosa* 30cm

Because of its ability to mimic, and its notes sounding close to the human voice, this bird is commonly kept in captivity, both in Thailand and throughout the world. It is glossy black, with large yellow wattles and a big orange bill. White wing patch in flight. Still fairly common in the wild in the forests of the north and west, and may be found in forests throughout the country up to about 1300m. Largely absent from the central plains and the north-east, and not normally found outside the forests. Resident.

BROWN-THROATED SUNBIRD *Anthreptes malacensis* 14cm

This fairly dull-coloured sunbird is one of the commonest sunbirds in the central plains and the peninsula. Like all sunbirds, it has a longish decurved bill, though nothing like as long as that of the spiderhunters. The male may be told by its light brown throat, iridescent greenish head, and purple shoulder patch. The female, like many female sunbirds, is olive above and yellow below, and has a dull yellow eye-ring. Very common resident in cultivated lowlands and the forest edge.

VAN HASSELT'S SUNBIRD *Leptocoma brasiliana* 10cm

With most male sunbirds, the plumage colours are often difficult to establish, depending on whether the bird is in direct sunlight or in shade. The male of this species has an iridescent green cap and a purple throat, dull red underparts, and a dark brown back with a purple sheen, but it often looks entirely black in the field. Uncommon forest-edge resident in the south-east and the peninsula. Within its range, the most confusable species is the Copper-throated Sunbird *Leptocoma calcostetha*, which lacks the purple throat.

OLIVE-BACKED SUNBIRD *Cinnyris jugularis* 11cm

This tiny olive and yellow sunbird has a blackish-looking throat and breast. The throat is in fact metallic blue-black, but this is reduced in winter to a small central stripe. The female has bright yellow throat and underparts and large white tips to the underside of the tail feathers. This species is found in open country, in deciduous woods, cultivated land and gardens, and in mangroves and coastal scrub in the south. Very common resident everywhere in the country.

GREEN-TAILED SUNBIRD *Aethopyga nipalensis* 11cm

An abundant resident on the upper slopes of Doi Inthanon, where it feeds at flowers almost down to ground level. The male has a dark green head and throat, a yellow breast band, a red lower breast, and yellow belly; the mantle is brown, and the rump yellow. The central tail feathers extend 3cm beyond the rest of the tail. Another race, resident in the peninsula, lacks red on the breast. Gould's Sunbird *Aethopyga gouldiae*, a winter visitor to the high mountains, lacks yellow on the breast and has more brighter red.

BLACK-THROATED SUNBIRD *Aethopyga saturata* 11cm

A rather dark sunbird, with extended central tail feathers as on Gould's and Green-tailed Sunbirds. It lacks the red colouring so conspicuous on males of those two species. The crown is metallic blue, the throat black, and the underparts pale yellow; the back is brownish, with the rump pale yellow. The race in the peninsula lacks yellow on the rump. Common resident in the evergreen forests of the north and a few other isolated areas throughout the country, between 300m and 1700m.

TEMMINCK'S SUNBIRD Aethopyga temminckii 11cm

The male is a tiny jewel: brilliant scarlet head, back, breast and tail, with yellow and purple on the lower back, and some slight purple markings on the head. Again, the elongated central tail feathers add a further 3cm to the total length. The female is much duller, with scarlet only on the feather edges of the wings and tail. A rare resident in evergreen forest up to 1500m, although most frequently found in the lowlands. Restricted to the southern parts of the peninsula.

LITTLE SPIDERHUNTER *Arachnothera longirostra* 16cm

A tiny spiderhunter with a bill almost half the length of its body. It is dark olive above and yellow below, with a white throat and very narrow black moustache, the latter bordered above by a diffuse area of white. The outer tail feathers are tipped white. This is characteristically a bird of forest undergrowth, commonest in or on the edge of dense evergreen forest. Common resident up to about 1000m throughout Thailand, except in the north-east and the central plains.

GREY-BREASTED SPIDERHUNTER *Arachnothera affinis* 18cm

This dull-coloured spiderhunter may be recognized by the absence of either yellow coloration or heavy streaking in its plumage. The upperparts are a rather dull olive-green, with blacker wings and tail, and the underparts are olive-tinged grey, only lightly streaked on the breast. Like all spiderhunters, it has a very long and decurved bill, but is otherwise a fairly nondescript bird. Common resident in the forests and secondary growth of the peninsula, up to 1100m.

STREAKED SPIDERHUNTER *Arachnothera magna* 19cm

This is the common spiderhunter of the northern hills. The very heavy streaking on both upperparts and underparts is diagnostic, and the pinky-orange legs and feet may also attract attention. It is often found near wild bananas, and is much easier to observe than the Little Spiderhunter, which occurs in the same range. Its presence is frequently first noticed from its loud chattering calls. Common resident up to 1800m in the north and west, in evergreen, mixed deciduous and secondary forest.

YELLOW-BREASTED FLOWERPECKER *Dicaeum maculatus* 10cm

A tiny bird with bright yellow underparts diffusely streaked with olive-green. The upperparts are deep olive, and it has a red patch on the crown. Juveniles lack the crown patch, and the streaks on the underparts are faint. The bill is thicker than that of most flowerpeckers. This species frequents the lower branches of the trees. A common resident of the southern half of the peninsula, where it inhabits forest and secondary growth up to 1600m.

YELLOW-VENTED FLOWERPECKER *Dicaeum chrysorrheum* 10cm

The conspicuous and well-defined dark streaking on the white underparts separates this from all other flowerpeckers. The yellow undertail-coverts are not always so easy to see. The upperparts are bright olive-green, with darker tail and wings. Prefers mixed forest, forest edge and clearings, and secondary growth, occurring from the plains up to about 1100m. Fairly common resident throughout the country, except in the north-east and the central plains.

ORANGE-BELLIED FLOWERPECKER *Dicaeum trigonostigma*
9cm

This is the only flowerpecker with deep blue-grey and orange plumage. The head is grey, with a paler throat and upper breast. The back, rump, and remainder of the underparts are bright orange. The wings and tail are dark grey. The female is noticeably duller and lacks the orange on the back, but retains a dull yellow-orange rump and yellowish underparts. Common resident in the southern half of the peninsula, in the majority of habitats, including wooded gardens, up to about 900m.

PLAIN FLOWERPECKER *Dicaeum concolor* 9cm

This tiny bird has no conspicuous markings and is difficult to distinguish from females of other species of flowerpecker. It is dull olive above and greyish below, perhaps somewhat brighter on the belly. The colouring of the head merges into the colour of the throat without any clear line of demarcation (female Fire-breasted Flowerpecker shows a clear dividing line). This species prefers the more open forests, as well as secondary growth. Fairly common resident, mostly in the north, up to 1700m.

SCARLET-BACKED FLOWERPECKER *Dicaeum cruentatum* 9cm

The brilliant scarlet of the crown and the entire upperparts of the male is quite unforgettable. The sides of the head, the wings and the tail are black, and the underparts white. The female is greyish-brown above, with a large scarlet rump patch. Juveniles lack red in the plumage but have a distinctive red bill base. This is the familiar flowerpecker of cultivated areas and gardens, but also extends to the deciduous forests and forest edge up to 1200m. Very common resident.

FIRE-BREASTED FLOWERPECKER *Dicaeum ignipectus* 9cm

This species replaces the Scarlet-backed Flowerpecker at the higher elevations of mountains in the north and south-east. The male is dark glossy blue-green above and pale yellow-buff below, with a scarlet patch on the breast (lacking on birds from the south-east) and a dark line down the belly centre. The female is dark olive-brown above and buffy below (*see* Plain Flowerpecker for differences). Very common resident in evergreen forest and areas of secondary growth above 600m.

JAPANESE WHITE-EYE *Zosterops japonicus* 12cm

The broad white ring around the eye distinguishes white-eyes from most other species (a few warblers also have this feature). This species is the common white-eye of the higher mountains, distinguished from the Oriental by its duller olive-green upperparts with no tinge of yellow. It has a yellow throat and upper breast, but no yellow on the central belly. Found in forests and cultivated areas. A common winter visitor to the north and parts of the north-east.

ORIENTAL WHITE-EYE *Zosterops palpebrosus* 11cm

Bright yellow-green above and yellow and grey below. The yellow of the breast continues down the centre of the belly, bordered by grey flanks; some birds are entirely yellow below. Wooded and garden habitats of all kinds. Very common resident in the north and in coastal areas of the south, ranging up to 1800m, though commonest at lower elevations. Two similar species are common winter visitors in the north: Japanese White-eye; and Chestnut-flanked White-eye *Zosterops erythropleurus* (small chestnut patch on the flanks).

EURASIAN TREE SPARROW *Passer montanus* 15cm

This familiar little bird of built-up areas needs no introduction. It is rich brown above, streaked with black; the crown is deep chocolate-brown, the cheeks white with a black patch, and the throat black. The underparts are dull brownish. Those living in cities usually look dull and drab. Sexes alike. Resident, and perhaps the most abundant bird of built-up areas and farmlands throughout the country, commonest in the lowlands, but extending up to 1800m on the mountains.

PLAIN-BACKED SPARROW *Passer flaveolus* 15cm

A relatively bright-plumaged sparrow. This species has no streaking on the back, which is plain chestnut (male) or light brown (female). The male also has chestnut behind the eye and continuing onto the shoulders, a black bib, and yellowish cheeks and underparts. The female lacks the bib, and the yellow is duller. Resident in lowland cultivated areas throughout the country. It is generally much scarcer than the Tree Sparrow, and tends to avoid urban environments.

BAYA WEAVER *Ploceus philippinus* 15cm

Male

Female

Weavers are thick-billed seed-eating birds of open country, preferably marshy with reedbeds. The males all have bright yellow crowns in breeding plumage. The male Baya Weaver is distinguished by its unstreaked tawny-buff underparts. Face and throat black; upperparts brown and streaked. Nests colonially, the woven nests often in full view; the nest of this species has a long entrance tube. Mainly in cultivated areas up to 1200m. Rather common resident in most parts of the country.

RED AVADAVAT *Amandava amandava* 10cm

An old name for this species, Strawberry Finch, is quite descriptive. The male is bright red, blacker on wings and tail, with white spots on the wings and underparts. Females, non-breeding males and young birds lack most of the red, but still have red bills. This species likes open grassy places and scrub, mainly in the lowlands, where small flocks may be found feeding in the tall grass. Easy to overlook! Uncommon resident in the central plains and the extreme north.

SCALY-BREASTED MUNIA *Lonchura punctulata* 11cm

This species often occurs in huge flocks in seeding bamboo and on rice paddies. It is plain brown above, often with somewhat paler uppertail-coverts; it has a brown throat, and distinctive scaly markings on the breast and flanks, with belly and undertail-coverts creamy-white. Juveniles lack the scaly markings, and have the throat and entire underparts buffish. Found in paddyfields and other cultivated land, also scrub, up to 1500m. Very common resident throughout the country except in parts of the peninsula.

CHESTNUT MUNIA *Lonchura atricapilla* 11cm

A small, rich chestnut bird with a black head and breast. The bill is blue-grey and, because of its colour, looks extremely heavy. Juveniles are a richer reddish-brown than juveniles of other munia species, and also have the same blue-grey bill as the adults. Found in grasslands, lowland marshes and scrub, paddyfields and coastal areas. An uncommon resident, found in several well-separated areas throughout the country, but absent from the north-east.

INDEX